AF731172

Tröge · Competition in Credit Markets

Beiträge zur betriebswirtschaftlichen Forschung

Schriftenreihe herausgegeben von:

Prof. Dr. Dr. h.c. mult. Horst Albach, Bonn
Prof. Dr. Sönke Albers, Kiel
Prof. Dr. Dr. h.c. Herbert Hax, Köln
Prof. Dr. Bernhard Pellens, Bochum

Band 94

Die „braune Reihe", wie die „Beiträge zur betriebswirtschaftlichen Forschung" häufig kurz genannt werden, ist eine der bekanntesten und angesehensten Buchreihen ihres Fachs. Seit 1954 erscheinen hier besonders qualifizierte, oft richtungsweisende Forschungsarbeiten (vor allem Dissertationen und Habilitationsschriften) der jeweils „neuen Generation" der Betriebswirtschaftslehre.

Michael Tröge

Competition in Credit Markets

A theoretic analysis

Springer Fachmedien Wiesbaden GmbH

Die Deutsche Bibliothek - CIP-Einheitsaufnahme

Tröge, Michael:
Competition in credit markets : a theoretic analysis
/ Michael Tröge. - 1. Aufl.. - Wiesbaden : Dt. Univ.-Verl. ; Wiesbaden : Gabler, 2001
(Beiträge zur betriebswirtschaftlichen Forschung ; Bd. 94)
Zugl.: Berlin, Humboldt-Univ., Diss., 1999
ISBN 978-3-8244-9048-6 ISBN 978-3-663-08310-8 (eBook)
DOI 10.1007/978-3-663-08310-8

1. Auflage September 2001

Alle Rechte vorbehalten
© Springer Fachmedien Wiesbaden 2001
Ursprünglich erschienen bei Betriebswirtschaftlicher Verlag Dr. Th. Gabler GmbH, Wiesbaden, und Deutscher Universitäts-Verlag GmbH, Wiesbaden, 2001

Lektorat: Ute Wrasmann / Dr. Tatjana Rollnik-Manke

www.gabler.de
www.duv.de

Das Werk einschließlich aller seiner Teile ist urheberrechtlich geschützt. Jede Verwertung außerhalb der engen Grenzen des Urheberrechtsgesetzes ist ohne Zustimmung des Verlages unzulässig und strafbar. Das gilt insbesondere für Vervielfältigungen, Übersetzungen, Mikroverfilmungen und die Einspeicherung und Verarbeitung in elektronischen Systemen.

Die Wiedergabe von Gebrauchsnamen, Handelsnamen, Warenbezeichnungen usw. in diesem Werk berechtigt auch ohne besondere Kennzeichnung nicht zu der Annahme, dass solche Namen im Sinne der Warenzeichen- und Markenschutz-Gesetzgebung als frei zu betrachten wären und daher von jedermann benutzt werden dürften.

Gedruckt auf säurefreiem und chlorfrei gebleichtem Papier.

ISBN 978-3-8244-9048-6

Preface

Since the 1960's, the foundations of finance have either been pure arbitrage or principal agent relationships. This study leaves the Modigliani - Miller world of pure arbitrage - but not the world of equilibria - and turns the principal-agent world upside down. Here the firms become principals again and banks become agents in the original sense of the word: they act on an auction market.

This new view of credit relationships yields a number of interesting insights. In my opinion the most important result is that too close relationships between banks and their borrowers will reduce credit market competition. Michael Tröge thus gives an antitrust reason for the limitation of bank involvement in non financial firms. This is not a very relevant issue in the United States where legal responsibility already makes it difficult for banks to influence the decisions of the firm. However, in continental Europe, close relationships between banks and firms are widespread and its effects on firms are subject to a large debate.

The author investigates in a first step the impact of the banks' information about borrower quality on the competitiveness of the credit market. Precise information about the credit quality can yield a competitive advantage for the bank but it does not come without cost. Tröge endogenizes the amount of information acquired in a strategic duopoly and obtains two key results: He first shows that too many banks can lead to excessive spending on information acquisition. Overbanking can therefore be socially harmful, a situation that will be especially likely to occur in economies with stable firms and anonymous credit markets.

His second and probably more important result is that under very general circumstances banks will specialize on firms and that this specialization will increase the cost of finance for the firm. The intuition is simple: Well informed banks can earn monopolistic rents. However as information acquisition is costly, it is not profitable for outsiders to acquire information about the firm and destroy the insider's rents.

Whereas these results suppose that the competitors cannot observe credit offers, Tröge also analyzes the situation in which bids are open - to a surprising effect. Whereas for a low number of bidders the intuition from the sealed bid case also applies to open bids, Tröge shows that for a high number of bidders additional effects arise: the value of information becomes negative for an inside bank. Banks prefer not to posses information about the firm or at least they prefer their competitor not to know about it.

In a forth chapter Tröge uses an the auction approach to generate insights regarding another important issue: Whether banks should be allowed to own equity in non financial firms. This question is be addressed from an antirust point of view. Share owning banks are remunerated by interest payments as well as by the firm's cash flows through dividends and increased firm value. Share owning banks therefore have an advantage in the competition for the credits to the firm and decrease the overall competitiveness of the credit market. Bank ownership of equity leads to higher interest rates that harm shareholders. Tröge concludes that restricting bank ownership in firms is crucial for ensuring of credit market competition.

Finally Tröge investigates auction markets where the company is still the principal, but the bank as an agent is able to ration credit. Remarkably the credit rationing is not necessarily detrimental to the firm looking for finance. The rationing bank will ration credit in order to signal to potential clients that it is not going to extract informational rents from the firm. This increases its reputation for not exploiting its clients. As a consequence agency costs are reduced.

The formal auction theoretic methods Tröge is develop employing are simple in construction but innovative and far from easy to solve. In this respect the work of Tröge is also an interesting contribution to pure auction theory.

Michael Tröge's research will deepen our understanding of relationship banking and bank finance. His results will draw attention to the important issue of competitiveness in credit markets. The time has come to rethink the relations between the house-bank system and competition in the credit markets and to implement the consequences for competition policy.

Horst Albach

Acknowledgments

This work would not have been possible without the help and assistance I have received from many sides.

I am especially indebted to Horst Albach who encouraged me to do research in the field of financial economics and provided me with the ideal working conditions by bringing me to the WZB. I am deeply grateful for his guidance and advice.

I would also like to express my thanks to Elmar Wolfstetter, who has given generously of his time to discuss about my work.

Among my colleagues and friends form the WZB and the Graduiertenkolleg "Angewandte Mikroökonomik" at the Humboldt University Rabah Amir, Ida Bagel, Uwe Dulleck, Hans Friederiszik, Dietmar Harhoff, Jim Y. Jin, Petra Korndörfer, Tobias Miarka, Silke Neubauer, Bill Novshek, Magdalena Paradowska-Thimm, Christoph Schenk and Lynda Thomas have all made comments and suggestions which have contributed substantially to this work. I owe particular thanks to Manfred Fleischer for his constant help and encouragement.

My research benefited a lot from discussions with participants at the EARIE98 and ESEM98 conferences, the annual meetings of the Deutsche Gesellschaft für Finanzierungswissenschaft and the Verein für Socialpolitik, the meetings of the DFG-Schwerpunkt "Inputmärkte"and the WZB conference on "Banking Competition and Financial Contracts" as well as from the seminar participants at the Humboldt University and the WZB. I would especially like to thank Sudipto Bhattacharya, Mark Flannery, Martin Hellwig and Ulrich Kamecke for their comments.

Michael Tröge

Contents

Summary

In this study, game-theoretic models for the strategic interaction of banks in the credit market are developed.

The first chapter motivates the analysis of credit market competition and gives a short overview of the existing literature and the questions that have been addressed.

In the second chapter, the auction approach to banking competition is introduced. The standard sealed-bid common-value framework is generalized by analyzing the banks' effort to acquire information about the firm before offering a credit. Two results are derived:

If banks can observe the others' effort to acquire information about a firm, asymmetric information between banks endogenously arises. One bank will then acquire a lot of information and become an insider or a housebank, whereas the other bank will stay uninformed. Hence exclusive housebank relationships endogenously arise in a narrow market where banks have a good knowledge about their competitors. It can be shown that this will reduce competitiveness and welfare.

The second result concerns relationship between the number of banks in a market, the quality of the firms and welfare. It is shown that, if the cost of information acquisition is taken into account, welfare and the firm's profit may decrease with a higher number of banks. In general a high number of banks produces more welfare with a bad sample of firms, whereas a low number of banks is preferable for a sample with a lot of good firms.

In the third chapter, the assumption of sealed bids is dropped. It is argued that, at least for the credits of large firms, open bid auctions are a more appropriate competition model. Similar to the closed bid setting, for a low number of bidders, informational rents can be sustained in an open bid auction.

However, whereas better information can be a competitive advantage in narrow markets, in markets with a high number of banks, good information about the quality of the firm may decrease the profit of the insider. With rising competition, low refinancing costs become a crucial success factor. This may explain why relationship banking is not compatible with intense competition.

In the forth chapter, the auction approach is used to analyze more precisely the implications of bank participation in the equity of a firm on the competitiveness of the credit market. Restrictions of equity ownership such as the National Banking Act and the Glass-Steagall–Act are part of the prudential regulation of banks in a large number of countries.

It is shown that these restrictions can be justified with antitrust arguments. An equity stake in a firm changes the payoffs of a bank in a way that is increasing its market power. The share-owning bank provides credit to this firm more often. The bank's profit and the average interest rate increases. If several banks owned symmetric stakes, the interest rate would decrease. However, if one bank owns an equity stake, no other bank will have incentives to own equity of this firm.

The fifth chapter analyzes the repeated interaction between a bank and a firm. A simple two period model is constructed which explains several puzzles: It shows why bank finance is available for firms which cannot obtain bond financing, why credit contracts contain a "Material Adverse Clause" and why interest rates quoted by banks do not depend very much on risk. The model shows why, even if the interest rate is observed, other banks cannot take over the credit.

In the second part of the chapter this model is used to give a new explanation for credit rationing. It is argued that banks may have to maintain a reputation for treating firms correctly. They will be reluctant to finance risky firms, because these credits will have to be renegotiated with a high probability which will endanger the bank's reputation. Rationing of this type can arise either in stable economies with a lot of low risk firms or in a highly risky environment. Non profit-maximizing public banks may then be necessary in order to subsidize small or risky firms.

List of Figures

Chapter 1

Introduction

1.1 Efficiency in financial markets

Efficient financial markets have to carry out several functions. Similar to other markets, efficiency requires that capital is provided at a price close to marginal costs. However, unlike in usual product markets, the allocation of capital is not only determined by its price. Banks choose their customers and play therefore an active role in allocating savings among alternative investment uses. Welfare is only maximized if they direct funds to the firms with the highest returns.

In addition, the provision of finance is not a spot transaction but extends over a period of time during which the provider of finance can influence the borrower. Again efficiency requires that the provider of finance finds ways to ensure that funds are not squandered.

Hence financial markets have to fulfill at least three tasks:

They should provide cheap finance, they should allocate the funds to the firms with the highest returns and they should provide efficient corporate governance[1].

The institutional arrangements which have evolved to provide these functions differ considerably between countries. Striking differences can for example be observed in the ratio of stock market capitalization to GNP which stretches from 152% in the United Kingdom to 27% in Germany (Monopolkommission (1998)). These differences in the size of the stock markets have led to the distinction between bank-based and market-based financial systems. However, the variety of financial systems should not only be reduced to these categories. There are other differences, not necessarily correlated with the stock market capitalization, which may have equally important consequences. Examples are the structure of the banking sector and the legal environment, in particular the regulation of banks or bankruptcy laws.

[1] These are the main functions which the financial system carries out for firms. Other tasks of financial intermediaries include: asset transformation, allocation of risk, liquidity provision, mobilisation of deposits, brokerage, maintenance of the payment system.

The literature emphasizing the importance of the financial system for the evolution of an economy dates at least back to Schumpeter (1911). However, until recently this literature did not have the formal arguments and elegance of the Arrow (1964), Debreu (1959) and Modigliani/Miller (1958) models, where finance was considered to have no importance.

Gerschenkron (1962) argues that initial stages of economic development can be better promoted by banking systems and controlled capital markets, whereas subsequent economic progress may benefit from more liberal securities markets. Allen (1993) suggests that banks may be more suited to monitor traditional and stable industries, whereas stock markets might be better at aggregating information about innovative technologies. Similar concerns have been raised by policy makers. For example the German Monopolkommission (1998) expresses strong concerns about the availability of finance for innovative firms due to the bank dominance of the financial system. The drawbacks of market-based systems have for example been pointed out by authors like Porter (1992). He argues that the predominant stock market financing in the US leads to a myopic concentration on short-term returns.

Empirical evidence that finance influences the economy is provided by a literature relating financial development and economic growth. In a early study, Goldsmith (1969) had observed a strong correlation between the development of the economy and the financial sector in a cross country study. However, he was not able to establish a causal relationship between the two. More recently, several papers have reexamined this empirical connection with new data and refined econometric methods (King and Levine (1993), Levine and Zervos (1998), Rajan and Zingales (1998), Rousseau and Wachtel (1998)). All papers provide strong evidence that it is not finance which reacts to the evolution of the real sector, but indeed the financial development which causes the differences in growth. Jayaratne et al. (1996) are able to demonstrate that even the finetuning of the banking regulation can be decisive. They examine the liberalization of the banking sector in different states in the United States and show that this had a positive influence on a state's growth. Enlarging the focus, Carlin and Mayer (1998) collect evidence that different financial systems not only influence overall growth, but are suited to different types of industrial activities and different stages of developments of firms and economies. They show that market centered systems, but also strong legal investor protection, are positively related to skill intensive industries. These observations suggest that small institutional or regulatory differences may have strong consequences. They may change the overall growth and direct the evolution of the economy in one or another development path.

However, the possibilities of empirical studies are necessarily limited. As financial systems differ by a wide variety of factors which may all influence one or several of the three components characterizing efficiency, it is almost impossible to empirically single out causalities or even trade-offs. Therefore, theoretical research is the only possibility to learn more about the complex relationship between the organization of financial markets and real economic activity. The relevant literature covers a wide range of topics along the causality chain stretching from depositors or investors over banks and stock markets to firms and product markets. One of the most important parts of this chain are banks. As Mayer (1988) has pointed out, independently of the structure of the financial system, banks are the most important source of external finance in all main capitalist industrialized nations. Nevertheless the strategic interaction of banks is not very well understood.

It is evident that a certain extent of competition between banks is necessary to achieve efficient financial markets. However, strong competition between banks has not always been regarded as a good thing. The main concern has been that excessive competition endangers the stability of the financial system. As the financial system has the character of an infrastructure, the failure of banks can have severe consequences for an economy. After experiencing a large number of banking failures already in the 19th century and again in the twenties and thirties of this century, most countries followed a policy of stabilizing the banking sector by limiting bank competition. Interest rates and fees were regulated or banks were allowed to fix them in cartel agreements, entry and the scope of action for banks was restricted. As the welfare losses caused by these restrictions became apparent, most of these measures were gradually relaxed during the last three decades (OECD 1989).

However, stability is not the only reason why competition could be harmful. Hellmann, Murdock and Stiglitz (1995a, 1995b) argue that too intense competition may lead to an underinvestment in deposit mobilization, inefficient allocation of resources and insufficient incentives to monitor firms and argue in favor of entry restriction and (temporary) deposit rate regulation. Carminal and Matutes (1997) show that competition may erode the incentives of banks to engage in monitoring and provide efficient corporate governance.

1.2 The scope of this study

In this study, basic game theoretical models of the strategic interaction of banks[2] in the credit market are developed. The focus is the impact of bank competition on firms. In particular, the relationship between competition and the three components of efficient financial markets (low cost of finance, efficient allocation and monitoring) will be established. Two very general results arise from the detailed analysis of the strategic situation: 1. Credit market competition is fundamentally different from competition in product markets. 2. Competition is heterogenous, the precise setup is very important, there is no single model which is appropriate for all situations.

[2]For simplicity, we will refer to all credit providing institutions as banks.

The main intention of this study is to provide a microeconomic analysis of oligopolistic bank competition which helps to understand the macroeconomic implications of bank dominated financial systems. A better understanding of the mechanism of banking competition is necessary to improve the regulation of banks, develop antitrust policy adapted to banking, evaluate the effect of bank mergers and anticipate the consequences of financial liberalization. This is especially important in the light of the current evolution of the financial systems. Globalization, deregulation and the progress of information processing technology have started to change the financial markets fundamentally. While the institutional environment undergoes major changes, the basic mechanisms of competition stay the same.

However, the game theoretic models that have been developed can be used in a multitude of other ways. In particular, the analysis could help banks to better adapt their business strategies to a changing competitive environment. And last but not least the models presented in chapter 2 and 3 can also be read as contribution to the general auction literature. The strategic situations described with these models may arise in a wide range of circumstances. The study identifies several effects which have not yet been described in the theoretical auction literature.

It has to be mentioned that a range of issues which are central to banking will be neglected. In particular, the question of financial stability will not be addressed at all. In addition, all questions which would involve depositors or investors are completely ignored. Therefore, important issues such as optimal risk sharing, deposit mobilization or liquidity provision cannot be addressed. The models in this study are at a less abstract level than much of the theoretical literature. The existence of banks and credit contracts is not explained, but assumed as given. For the purposes of this study, it is sufficient to motivate the existence of banks with arguments given in the seminal paper of Diamond (1984), the form of the credit contracts may be explained following Innes (1990).

1.3 Competition between banks

1.3.1 The industrial organization approach

Bank representatives like to claim that money is a very homogenous good and that therefore the credit market is necessarily perfectly competitive. Indeed most of the banking literature uses the Bertrand model with identical marginal costs (refinancing rates) where banks do not make any profits. However, banks clearly make profits. One of the earliest models of bank behavior, the Monti-Klein model (Klein 1971, Monti 1972) avoids this paradox by assuming monopolistic banks. A more realistic model of banking competition, which also generates nonzero rents, is the generalization of the Monti-Klein model to a two sided Cournot oligopoly (compare Rochet/Freixas (1997 p. 59). In fact, Yanelle (1989) and (1997) has shown that nonzero rents and inefficiency can also be the result of two-sided Bertrand competition, where banks first compete for deposits and then on the credit side.

Another possibility to avoid the Bertrand paradox which has become popular in the banking literature is the Salop (1979) spacial competition model. It has been used for modelling competition on the deposit side (Matutes and Padilla (1994), Cordella and Levy-Yeyati (1998)) but also on the credit side (Chiappori et al. (1995), Gehrig (1996)). Indeed, the Salop model is probably a good description for what happens on the retail and depository side of the business. Customers have to pass frequently at a bank for example to get money and tend therefore to choose a bank close to their home. However, product differentiation is probably not a good explanation of the economic mechanism that allows a bank to earn rents in the credit business. Normally, banks are sufficiently close to each other for the travelling costs to be negligeable compared to the amount of credit. It has been argued that differentiation should be understood in a more abstract sense for example as differences in services provided by the bank. However, services provided by the banks are rather a different product than a property of the credit and should therefore be priced and analyzed separately.

1.3.2 The auction approach

This study argues that it is misleading to consider banks, in analogy to usual firms as producers of credit and borrowers as consumers. There is a simple reason: In most markets, producers doe not care about who buys the product[3], whereas banks care very much to whom they give the credits. As a credit is an exchange of money against money, it is not evident what is the price and what the good. It may be a more appropriate way of looking at banking to exchange the roles of the seller and the buyer. In fact, the credit can be understood as the price a bank pays now to obtain the right to an uncertain future cash flow. This is an approach closer to the theory of financial markets, where the firm sells bonds to the investors. Unlike in financial markets, however, as usually there are only a few banks, we cannot reasonably model the buyers as price takers. When several banks compete for a loan, the price is determined through bids submitted by the banks, i.e. in an auction[4].

In the auction setting, the banks' rents do not arise from product differentiation in the usual sense. They are informational rents caused by the informational asymmetries between banks. Two types of information, corresponding to the allocation and the corporate governance function of a bank, can be distinguished.

[3]Exceptions are reputational effects of the buyers identity on the producer.

[4]To be precise, in a credit auction, bids are not in prices (i.e. sums payed by the bank to the firm) but, by announcing interest rates, banks determine what fraction of the object of uncertain value they want to have in exchange for the fixed price (loan) they pay.

Before financing a firm, the bank needs information about the quality of the borrower in order to avoid financing unprofitable firms. The collection of information before offering a credit has been called "ex ante monitoring" or "screening". Banks rely on this information in order to decide about credit approval. Therefore sufficient screening is crucial for the allocation of capital in an economy. If banks have insufficient incentives to screen, they will refuse to finance complicated or innovative businesses the creditworthiness of which is difficult to judge. Especially freeriding and spillover effects may prevent the bank from earning the marginal rent of its information collecting effort and erode the bank's screening effort.

However, if the bank has already financed the firm, information about the quality is only of limited use. Then the bank needs information which enables it to decrease the probability of credit losses, i.e. to provide corporate governance. The kind of information that will be useful depends on the possibilities the bank has to influence a firm. If it can only withdraw its credit or increase the collateral, it will be sufficient to know the firm's business well enough to anticipate distress situations early. If the bank has more extensive channels of influence, for example through a post in the supervisory council, it will be useful to develop knowledge of how to conduct the firm's business, influence personnel decisions etc.. The acquisition of information destinated to improve corporate governance has been referred to as "monitoring".

Screening The auction approach to banking has been initiated by Broecker (1989). He models bank competition as a sealed bid common value auction where banks get an imperfect binary signal about the quality of the firm. The key insight of his paper is that if banks have independent imperfect information about the quality of a borrower, the screening activity of one bank has a negative impact on the other's profit. The firms refused by one bank will apply for credit at the other bank and reduce the average quality of this bank's loan applicants. In a model with binary signals, the competition of banks in interest rates has no pure strategy equilibrium. All firms that have been accepted by both banks will take the credit with the lower interest rate. Only firms, which have been refused by the bank quoting low interest rates, will go to the bank offering the higher interest rate. The average quality of these firms will be so bad that, despite the higher interest rate, this bank will make lower profits. It is therefore always profitable to slightly underbid the other bank and receive the good borrowers.

In Broecker's paper, the quality of the signal received by the firms is completely exogenous. A step further in the direction of endogenizing the ex ante information acquisition by banks has been made by Riordan (1993). He also models bank competition a sealed bid common value auction, but, in contrast to Broecker's model, banks receive a continuous signal of the borrower's quality. They are therefore able to quote a different interest rate for each borrower and the resulting equilibrium is in pure strategies. Riordan analyzes how the interest rates and loan approval decisions of each bank depend on the number of competitors and how the number of loans approved by the market changes with the market structure. However, he doesn't endogenize the effort deployed by banks to acquire information. The informativeness of the creditworthiness tests is still exogenous, Riordan only determines endogenously how the bank uses this information.

Kanniainen and Stenbacka (1998) try to completely endogenize the screening effort and analyze the socially inefficient distortions of the banks screening effort caused by competition. In a similar setting, Gehrig shows that if information acquisition is taken into account, monopolistic outcomes are stable with respect to competition. Unfortunately, both papers suffer from technical flaws. Indeed, endogenizing the information acquisition of bidders is still an unsolved problem in general auction theory. Matthews (1985) has formulated an auction with bidders of asymmetric information and endogenised the information acquisition in the context of pure common values and closed bidding. Despite assuming a quite simplified information acquisition technology, he is not able to ensure the existence of an equilibrium. One way to get around the mathematical difficulties is to assume a discrete information acquisition technology and solve the mixed equilibrium strategies. This is how Hausch and Li (1993) proceed. Ruckes (1998) has adapted this method for modeling banking competition. He is able to show that competing banks will change their credit standards with the quality of the firms in the market. If there are a lot of good firms they will be much more lenient than if the risk of credit losses is high.

Other papers have avoided to model explicitly competition in screening and focussed on how screening will be influenced by other variables. Chan, Greenbaum and Thakor (1986) show that the screening effort of banks decreases with the reusability of information, hence will be low in fast changing, risky environments. Manove, Padilla and Pagano (1998) show that the presence of collateral may lead to an inefficiently low screening level. In addition, banks will not acquire the socially relevant information. For example, it is to be suspected that they will be more interested to know the default probability than the maximal return in a best case scenario.

Monitoring and the lock-in effect The idea that a bank may acquire information during a lending relationship has been pioneered by Sharpe (1990). He constructs a multiperiod model of a lending relationship where, during the first period, the banks gain inside information about the quality of the firm. It is assumed that this information gives the bank an informational advantage which prevents other banks to enter in competition at a later stage. The inside banks have therefore the possibility to extract ex post monopolistic rents. However, as banks are perfectly competitive at the outset, these rents are competed away. In the first period banks quote interest rates below their marginal costs in order to attract customers. Hence, the interest rates offered to firms will be low at the beginning of a relationship and increase afterwards. In Sharpe's model these interest fluctuations lead to inefficient investment by the firms. Sharpe demonstrates how in a simple infinitely repeated game, banks can acquire reputation not to exploit their informational advantage at the borrower's expense and implement socially optimal investment.

Whereas previous theoretical studies had always pointed out the beneficial effects of bank relationships in reducing agency costs, the ex post rent-extraction effect identified by Sharpe was one of the first rigorous arguments why bank relationships could also have drawbacks and has therefore attracted considerable attention. Using an auction model, Fischer (1990) studied the competition of an informed and an uninformed bank in detail and pointed out that Sharpe was not entirely correct in assuming that privileged information about a borrower's quality prevents competition. In fact, the strategic situation will result in an equilibrium with mixed strategies where insiders will earn higher rents, but outsiders will sometimes obtain the credit. Corollary 4 in chapter 2 obtains his result as a special case of a more general game. The same competition model was used in Rajan's (1992) analysis of the consequences of ex post rent-extraction. He compares the incentives for the firm provided by informed versus arm's length finance. Bank finance reduces the possibility of the entrepreneur to increase the risk of his project, but also reduces the entrepreneur's incentives to invest effort, because he anticipates rent extraction. Rajan concludes that borrowers who anticipate a sequence of very profitable projects would prefer arm's length or market financing to bank finance.

Rajan (1992) and Sharpe (1990) assume that inside banks can extract rents, because they have a better information about the borrower's credit risk. In chapter 3 of this study, it is shown that this effect may disappear if outsiders can observe the interest rates offered by the insider. In chapter 5, a different mechanism is described which allows the bank to extract ex-post monopolistic rents even if the interest rate is observable. It is argued that the advantage of the inside bank cannot be explained by a better knowledge of the borrower's quality, but is due to the fact that inside banks can provide better corporate governance by anticipating distress situations. A related approach has been proposed by Garella (1996) who supposes that the rents generated by banks can be explained by the bank's reduction of a borrowers riskiness. Similar to Rajan (1995), in chapter 5 it is endogenously motivated why banks acquire information about the firm and not simply assumed that bilaterally available proprietary information is produced spontaneously during a bank-borrower relationship.

Chapter 2

Information acquisition in credit auctions

2.1 Introduction

In this chapter, the incentives to acquire costly information in a sealed bid common value model of credit market competition will be analyzed. The existing work is generalized in two directions:

In the first part of the chapter, a duopoly is considered, in which the quality of both banks' information can be varied continuously and independently. The model includes therefore the Ruckes (1998) duopoly model, where banks have information of symmetric quality, as well as the Fischer (1990)/Rajan (1992) models, where a perfectly informed bank competes against a completely uninformed one. Allowing both banks to acquire information of different quality makes it possible to demonstrate an effect which will lead to decreased competitiveness in narrow markets, where banks have a good knowledge about the others' activities. It will be shown that, if the quality of the competitors' information can be observed by a bank, asymmetric information endogenously arises. One bank will acquire a lot of information and become an insider whereas the other bank has low incentives to engage in information acquisition and stays outsider. Compared to the welfare optimizing amount of information acquisition, the inside bank will overinvest and the outside bank will underinvest. Nevertheless, the profits of both insider and outsider will be higher than with the welfare optimal information acquisition.

In the second part of the chapter, an oligopolistic market with n banks is analyzed, under the assumption that banks cannot observe the quality of the competitors' information. In this case, every bank will invest the same effort in information acquisition and the equilibrium will be symmetric. It is shown that if the sample of firms is rather bad, the banks will acquire the welfare optimal amount of information. However, if there are a lot of firms in the sample, banks will overinvest in information acquisition. Therefore welfare may decrease with the number of banks. In contrast to usual competition models, more competition does not necessarily increase welfare.

Whereas the basic results of this chapter are intuitive, the precise calculation of the equilibria is technically demanding.

2.2 The model

The market consists of n banks competing to give a credit to one firm. The size of the loan is normalized to one. The firm has an investment project, which is going to succeed with probability λ. In this case the firm's return is X, whereas the unsuccessful projects return nothing and the firm goes bankrupt.

By taking a close look at the firm, for example by examining the books, analyzing the feasibility of the firm's project and evaluating the quality of the firm's staff, banks are able to get better information about whether the firm is going to be successful. However, this information will not come without cost.

This situation is modeled by assuming that banks are equipped with a costly information production technology. Investing γq^2 in information acquisition, a bank independently receives with probability q a perfect signal about the quality of the firm. This means that with probability q this bank will know with certainty if the firm is going to succeed or not, whereas with probability $1 - q$ the bank does not receive additional information. It is important to note that they do not know whether the competitors have received any information.

However, banks may or may not observe the competitor's investment in information acquisition, i.e. the quality of the competitor's information. In narrow markets with good relationships between the banks, it seems reasonable to assume that they know the probability q_j, with which the competitors receive information about a given firm. In this case, the choice of q_i has to be modeled as an independent first step of the game. Before the offers are made in the second step, q_i will be revealed and cannot be changed anymore. The investment has therefore commitment value. This is the approach taken in the first part of this chapter. It will be shown that this leads to asymmetric situations which are less competitive.

Alternatively, the investment in information acquisition and the decision about the offer can be modeled as simultaneous choices in a one step game with two decision variables. This is the more appropriate way of modelling anonymous markets with a lot of banks where it cannot be assumed that every bank knows how well informed the competitors are. In the second part of this chapter, it will be shown that this game leads to more competitive symmetric equilibria.

Depending on their information, the banks may offer the firm a credit, asking for the repayment of b in case of success. As the investment has been normalized to one, this corresponds to an interest rate of $b - 1$. The bidding is assumed to be closed i.e. banks do not know the competitor's bids nor the fact that the competitor has made an offer.

It is clear that whenever a bank has received the information that the firm is going to fail it will not make a credit offer. Accordingly when a bank knows that the project is going to succeed it will always offer credit. In case a bank has received no additional information it may or may not offer credit, depending on the ex ante probability of success of a firm. When deciding about the interest rate, a bank faces a trade-off between a higher profit in the case of winning and a higher probability of winning but a lower interest rate. Similar to other auctions with discrete values, this game has no equilibrium in pure strategies. If one bank were always bidding the same interest rate, the best response of the other bank would be either to slightly undercut this bid or to always bid the highest possible amount. Clearly, in both cases the first bank's bid is not optimal.

2.3 Asymmetric equilibrium with two bidders

In this section, the sequential version of the game is solved by backward induction. First the last step has to be solved. In this step the two banks have already received information with the probabilities q_1 and $q_2 \in (0,1)$. Without restriction of generality $q_1 \leq q_2$ is imposed.

An equilibrium in mixed strategies can be described by the bidding densities $F_j(b)$ of bank j in case it has received a good signal, the probability μ_j of making an offer in case the bank j has not received a signal and the distribution of the bids $H_j(b)$ in case the bank is bidding without having received a signal. The profit of a bank i, having received a good signal and bidding b, can be calculated as follows:

$$\pi_i^g(b) = (b-1)\left[q_j(1-F_j(b)) + (1-q_j)\left[1-\mu_j H_j(b)\right]\right], \quad i,j = 1,2 \tag{2.1}$$

This is the profit on the credit business $(b-1)$ multiplied by the probability of giving the loan. If bank i has received a good signal, the project must be successful. This means that, if the competitor j also receives a signal, this must be a good signal. He will therefore make an offer which, with probability $1-F_j(b)$, will be higher than the offer b made by bank i. With probability $1-q_j$, bank j will not receive a signal. In this case bank j will make an offer with probability μ_j. This offer will not be accepted with probability $1-H_j(b)$.

The profit of bank i, offering credit at an interest rate of $b-1$ despite not having received a signal, can be obtained with similar reasoning:

$$\pi_i^0(b) = -(1-\lambda)q_j + (b-1)\lambda q_j(1-F_j(b)) + (\lambda b - 1)(1-q_j)\left[1-\mu_j H_j(b)\right], \tag{2.2}$$

for $i,j = 1,2$. In a mixed strategy equilibrium, a bank must be indifferent between bids on the support of its bidding distribution. This condition leads to a set of four equations which, for $q_1 \neq q_2$, have a unique solution. The equilibrium is, however, quite complicated. The following definitions will facilitate the presentation:

Definition 1 *The bids b_1, b_2, the disjunct intervals $\mathcal{I}_1 \ldots \mathcal{I}_5$ in the bidding space, and the probabilities μ_1^b and μ_1^c are defined as follows:*

$$\begin{aligned} \mu_1^b &:= 1 - \frac{(1-\lambda)}{(1-q_1)(X-1)\lambda}, & \mu_1^c &:= 1 - \frac{(1-\lambda)q_2}{(1-q_1)(X\lambda - 1)}, \\ b_1 &:= \frac{1-\lambda q_1}{\lambda(1-q_1)}, & b_2 &:= \frac{1-\lambda q_2}{\lambda(1-q_2)}, \\ \mathcal{I}_1 &:= \left(-\infty, \frac{1}{\lambda}\right), & \mathcal{I}_2 &:= \left[\frac{1}{\lambda}, b_1\right), \\ \mathcal{I}_3 &:= [b_1, b_2), & \mathcal{I}_4 &:= [b_2, X) \text{ if } b_2 < X, \\ \mathcal{I}_5 &:= [X, \infty). \end{aligned} \tag{2.3}$$

The assumption $q_1 \leq q_2$ implies $b_1 < b_2$. Depending on whether $b_1 > X$, $b_1 < X < b_2$ or $X > b_2$, the equilibrium takes three qualitatively different forms.

Proposition 1 *(equilibrium strategies) If two banks are receiving information with probability q_1 and q_2, $q_1 \leq q_2$, the equilibrium is of the following form:*
a) If $X < b_1$, both bidders only bid if they have received a good signal. Their bids are distributed according to:

$$F_1(b) = \begin{cases} \frac{1}{q_1}\left[1 - (1-q_1)\frac{X-1}{b-1}\right] & \text{for } b \in [q_1 + X - q_1 X, X], \\ 0 & \text{for } b < q_1 + X - q_1 X, \\ 1 & \text{for } b > X, \end{cases} \tag{2.4}$$

$$F_2(b) = \begin{cases} \frac{1}{q_2}\left[1 - (1-q_1)\frac{X-1}{b-1}\right] & \text{for } b \in [q_1 + X - q_1 X, X], \\ 0 & \text{for } b < q_1 + X - q_1 X, \\ 1 & \text{for } b > X. \end{cases} \tag{2.5}$$

b) In the case $b_1 < X < b_2$, bank 1 bids with good signal and with probability μ_1^b having received uninformative signals, whereas bank 2 only bids with a good signal. The equilibrium bidding distributions are:

$$F_1(b) = \begin{cases} 0 & \text{for } b \in \mathcal{I}_1, \\ \frac{\lambda b - 1}{\lambda q_1 (b-1)} & \text{for } b \in \mathcal{I}_2, \\ 1 & \text{for } b > \mathcal{I}_3 \cup \mathcal{I}_4 \cup \mathcal{I}_5, \end{cases} \tag{2.6}$$

$$F_2(b) = \begin{cases} 0 & \text{for } b \in \mathcal{I}_1, \\ \frac{\lambda b - 1}{\lambda q_2 (b-1)} & \text{for } b \in \mathcal{I}_2 \cup \mathcal{I}_3 \backslash [X, \infty), \\ 1 & \text{for } b \in [X, \infty), \end{cases} \tag{2.7}$$

$$H_1(b) = \begin{cases} 0 & \text{for } b \in \mathcal{I}_1 \cup \mathcal{I}_2, \\ \frac{1}{\mu_1^b}\left[\frac{b}{(b-1)} - \frac{1-\lambda q_1}{(b-1)(1-q_1)\lambda}\right] & \text{for } b \in \mathcal{I}_3, \\ 1 & \text{for } b \in [X, \infty). \end{cases} \tag{2.8}$$

c) For $X > b_2$, both bidders bid with good as well as with inconclusive signals. Bank 2 will always bid, even in the case of an inconclusive signal, whereas bank 1 will only bid with probability μ_1^c after having received an inconclusive signal. The equilibrium bidding distributions are:

$$F_1(b) = \begin{cases} 0 & \text{for } b \in \mathcal{I}_1, \\ \frac{\lambda b - 1}{\lambda q_1 (b-1)} & \text{for } b \in \mathcal{I}_2, \\ 1 & \text{for } b > \mathcal{I}_3 \cup \mathcal{I}_4 \cup \mathcal{I}_5, \end{cases} \tag{2.9}$$

$$F_2(b) = \begin{cases} 0 & \text{for } b \in \mathcal{I}_1, \\ \frac{\lambda b - 1}{\lambda q_2 (b-1)} & \text{for } b \in \mathcal{I}_2 \cup \mathcal{I}_3, \\ 1 & \text{for } b > \mathcal{I}_4 \cup \mathcal{I}_5, \end{cases} \tag{2.10}$$

$$H_1(b) = \begin{cases} 0 & \text{for } b \in \mathcal{I}_1 \cup \mathcal{I}_2, \\ \frac{1}{\mu_1^c}\left[\frac{b}{(b-1)} - \frac{1-\lambda q_1}{(b-1)(1-q_1)\lambda}\right] & \text{for } b \in \mathcal{I}_3, \\ \frac{1}{\mu_1^c}\left[1 - \frac{(1-\lambda) q_2}{(b\lambda - 1)(1-q_1)}\right] & \text{for } b \in \mathcal{I}_4, \\ 1 & \text{for } b \in \mathcal{I}_5, \end{cases} \tag{2.11}$$

$$H_2(b) = \begin{cases} 0 & \text{for } b \in \mathcal{I}_1 \cup \mathcal{I}_2 \cup \mathcal{I}_3, \\ \frac{1}{1-q_2}\left[1 - \frac{b-1}{\lambda b - 1}\lambda q_2\right] & \text{for } b \in \mathcal{I}_4, \\ 1 & \text{for } b \in \mathcal{I}_5. \end{cases} \tag{2.12}$$

Proof. see Appendix A ■

The most relevant case is a). The condition $b_1 > X$ is equivalent to $\lambda < \frac{1}{q_1 + (1-q_1)X}$, which always holds if $X\lambda < 1$, i.e. if the sample of firms is so bad that, without additional information, it is ex ante not worthwhile to finance a firm. Even if $X\lambda > 1$, for high degrees of informativeness, $b_1 > X$ still holds.

Situation b) or c) where banks bid on inconclusive signals only occur if the sample of firms is very good, i.e. $\lambda X > 1$ and the information of the banks rather low.

In the case a) the equilibrium distribution functions do not depend on λ. Figure 2.1 shows the distribution function of both banks in this case for $q_1 = 0.6$,

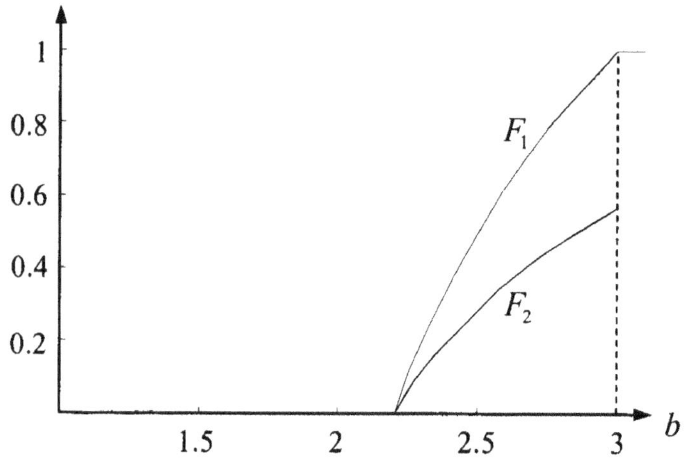

Figure 2.1: Case a): Bidding distributions with $q_1 = 0.6$, $q_2 = 0.8$ and $X = 3$

$q_2 = 0.8$ and $X = 3$. Both banks randomize on the same support. We have $F_2(X) = \frac{q_1}{q_2} < 1$. This means that the bank with the better information is bidding the highest possible bid with a positive probability. In expectation, it is also asking for a higher interest rate.

Figure 2.2 shows the bidding functions in the case b), for $\lambda = 0.54$, $q_1 = 0.4$, $q_2 = 0.7$, $X = 3$.

In this case, the less informed bank will bid when having received an inconclusive signal. However, it will make an expected profit of zero in these cases.

Figure 2.3 shows the situation c) for $q_1 = 0.4$, $q_2 = 0.6$, $\lambda = 0.6$, $X = 3$. In this case, the bank with the better information offers always a credit except if it knows that the firm is unsuccessful. It makes positive expected profit, even having received an inconclusive signal. The less informed bank only bids with probability μ_1 in case of an inconclusive signal and does not make an expected profit on these bids.

The equilibrium profits can be calculated by plugging back the equilibrium strategies 2.4...2.12 into the profit functions 2.1...2.2:

Proposition 2 *In case a), the expected profit in case of having received a good signal is for both banks* $\pi_i^g = (X-1)(1-q_1)$, $i = 1, 2$. *The overall profits are*

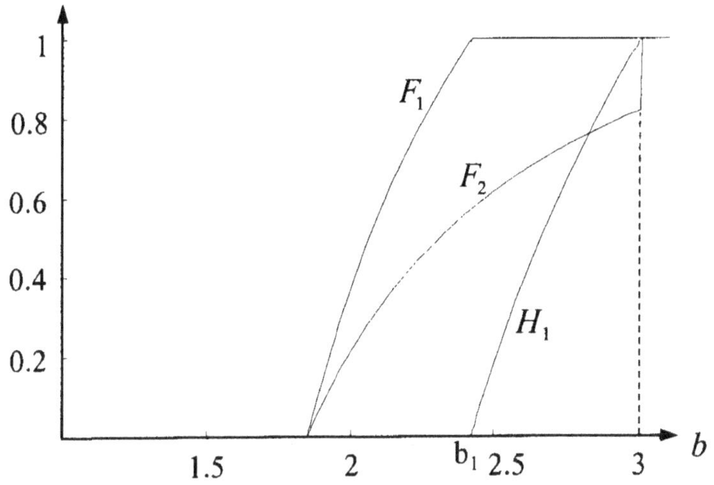

Figure 2.2: Case b): Bidding distributions for $\lambda = 0.54$, $q_1 = 0.4$, $q_2 = 0.7$, $X = 3$

therefore

$$\pi_2 = \lambda q_2 (X-1)(1-q_1), \tag{2.13}$$
$$\pi_1 = \lambda q_1 (X-1)(1-q_1). \tag{2.14}$$

In case b), both banks make a profit of $\pi_i^g = \frac{1}{\lambda} - 1$ having received a good signal and no profit with an inconclusive signal. Hence

$$\pi_i = q_i (1-\lambda) \quad i = 1,2. \tag{2.15}$$

In case c), the profit after a good signal is $\pi_i^g = \frac{1}{\lambda} - 1$. Bank two having received an inconclusive signal is making a profit of $\pi_2^0 = (q_2 - q_1)(1-\lambda)$, bank one is making no profit in case of an inconclusive signal. The overall profits are therefore

$$\pi_2 = [q_2 + (1-q_2)(q_2 - q_1)](1-\lambda), \tag{2.16}$$
$$\pi_1 = q_1 (1-\lambda). \tag{2.17}$$

Proof. *see Appendix A* ■

The case $p_1 = 0$, $p_2 = 1$ is especially interesting. This is the insider versus outsider situation, analyzed by Rajan and Fischer, where one bank is perfectly

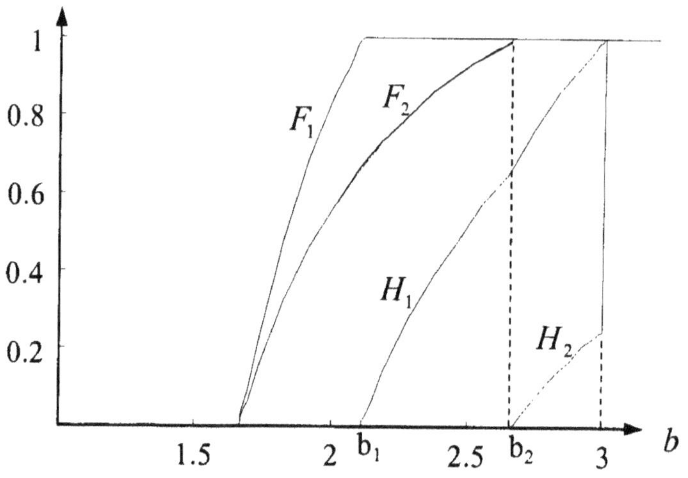

Figure 2.3: Case c): Bidding distributions for $q_1 = 0.4, q_2 = 0.6, \lambda = 0.6, X = 3$

informed about the quality of the firm, whereas the other bank has no information. Some of the intervals $\mathcal{I}_1 ... \mathcal{I}_4$ disappear, therefore the equilibrium is not precisely included in proposition 1. However, the equations 2.3, 2.7 and 2.8 still describe the equilibrium.

Corollary 1 *(Fischer/Rajan) If one bank knows the quality of the firm, whereas the other is uninformed, the equilibrium strategies are:*
a) For $\lambda X < 1$, the outsider is not bidding. The insider is bidding X in case he gets a good signal and obtains the entire surplus.
b) For $\lambda X > 1$, the outsider participates with probability $\mu_1^b = \frac{\lambda X - 1}{\lambda (X-1)}$, bidding:

$$H_1(b) = \begin{cases} 0 & \text{for } b < \frac{1}{\lambda}, \\ \dfrac{(\lambda b - 1)(X-1)}{(b-1)(\lambda X - 1)} & \text{for } b \in \left[\frac{1}{\lambda}, X\right), \\ 1 & \text{for } b \in [X, \infty). \end{cases} \tag{2.18}$$

The insider bids

$$F_2(b) = \begin{cases} 0 & \text{for } b < \frac{1}{\lambda}, \\ \dfrac{\lambda b - 1}{\lambda (b-1)} & \text{for } b \in \left[\frac{1}{\lambda}, X\right), \\ 1 & \text{for } b \in [X, \infty). \end{cases} \tag{2.19}$$

Proof. See Appendix A ■

The lack of information does not mean that the outsider will never win the auction, as assumed by Sharpe (1990). From 2.15, we see that the outsider does not make profits, but prevents the insider form earning more than an expected profit of $1 - \lambda$.

The symmetric equilibrium for two equally well informed banks with $q_1 = q_2$ has been analyzed by Ruckes (1998). However, the equilibrium described in his paper is not unique. Only in the case a) there is a unique equilibrium described by setting $q_1 = q_2 = q$ in equation 2.4. In case c), there is a continuum of payoff-equivalent but not welfare-equivalent equilibria:

Corollary 2 *(Ruckes) For*

$$\lambda > \frac{1}{q + (1-q)X} \tag{2.20}$$

there is a set of equilibria where, having received a good signal, both banks are bidding:.

$$F_i(b) = \begin{cases} 0 & \text{for } b \in \mathcal{I}_1, \\ \dfrac{\lambda b - 1}{\lambda q (b-1)} & \text{for } b \in \mathcal{I}_2, \\ 1 & \text{for } b > \mathcal{I}_3 \cup \mathcal{I}_4 \cup \mathcal{I}_5. \end{cases} \tag{2.21}$$

Having received no signal, one bank is bidding with probability $\mu_1 = 1 - \frac{(1-\lambda)q}{(1-q)(X\lambda - 1)}$, *using the distribution function:*

$$H_1(b) = \begin{cases} 0 & \text{for } b \in \mathcal{I}_1 \cup \mathcal{I}_2, \\ \dfrac{1}{\mu_1}\left[1 - \dfrac{(1-\lambda)q}{(b\lambda - 1)(1-q)}\right] & \text{for } b \in \mathcal{I}_4, \\ 1 & \text{for } b \in \mathcal{I}_5. \end{cases} \tag{2.22}$$

The other bank has then the choice to bid with any probability $\mu_2 \in \left(1 - \frac{(1-\lambda)q}{(X\lambda - 1)(1-q)}, 1\right]$ *and the corresponding distribution function:*

$$H_2(b) = \begin{cases} 0 & \text{for } b \in \mathcal{I}_1 \cup \mathcal{I}_2, \\ \dfrac{1}{\mu_2}\left[1 - \dfrac{(1-\lambda)q}{(b\lambda - 1)(1-q)}\right] & \text{for } b \in \mathcal{I}_4, \\ 1 & \text{for } b \in \mathcal{I}_5. \end{cases} \tag{2.23}$$

The profit of both banks is $\pi = q(1-\lambda)$.

Proof. see Appendix A ■

The banks will not make profit when bidding without having received a good signal. Therefore, they are indifferent between participating at the auction or not. One of them can bid the highest value with a positive probability and participate more often in the auction without changing the indifference condition for the other bank. This will not change its profit, but increase the probability of a firm getting finance, even if both banks have received an inconclusive signal and therefore increase social welfare.

2.3.1 Endogenous information acquisition

The profit of the banks depends on the quality of the own information as well as on the information of the competitors. In order to realize their profit in the second step, banks have to invest γq_i^2 in information acquisition. Using the profit formulas from proposition 2, their overall payoffs are therefore

$$\Pi_1(q_1, q_2) = \pi_1 - \gamma q_1^2, \quad (2.24)$$
$$\Pi_2(q_1, q_2) = \pi_2 - \gamma q_2^2. \quad (2.25)$$

Depending on how much information is acquired, the banks will be in one of the situation a) to c) from proposition 1.

Proposition 3 *a) For* $\gamma \leq \lambda \frac{2(X-1)-\lambda(X^2-1)}{\lambda X-1}$ *or* $\lambda X < 1$, *the bidders will acquire the amount of information*

$$q_1^* = \frac{(X-1)\lambda}{2(\gamma + (X-1)\lambda)}, \quad (2.26)$$

$$q_2^* = \min\left\{\frac{(X-1)\lambda}{2\gamma}\left[1 + \frac{(X-1)\lambda}{2(\gamma + (X-1)\lambda)}\right], 1\right\}. \quad (2.27)$$

b) For $\gamma > \lambda \frac{2(X-1)-\lambda(X^2-1)}{\lambda X-1}$ *and* $\lambda X > 1$,*the equilibrium levels of information will be:*

$$q_1^* = \frac{1-\lambda}{2\gamma}, \quad (2.28)$$
$$q_2^* = \frac{(1-\lambda+4\gamma)(1-\lambda)}{(1-\lambda+\gamma)4\gamma}. \quad (2.29)$$

Proof. See Appendix A ■

Again a) is the most relevant case. Remarkably even if the cost of information acquisition tends to zero the outsider will not acquire more information than necessary to receive a conclusive signal in half of the cases. In particular, there will always be an outsider and an insider.

2.3.2 Welfare

For simplicity the analysis is restricted to the most relevant case $\lambda X > 1$. In this case banks are only financing if they know the project is good. Therefore, no money will be lost in financing inefficient projects. In the model, high interest rates simply redistribute the surplus, but do not cause any welfare decreasing distortions of investment decisions. Therefore, welfare only depends on the probability of a project being financed:

$$W\left(q_2^*, q_1^*\right) = \lambda\left(X-1\right)\left[1-\left(1-q_1^*\right)\left(1-q_2^*\right)\right] - \gamma\left(q_1^{*2}+q_2^{*2}\right). \tag{2.30}$$

The following proposition compares the welfare optimal amount of information acquisition with the equilibrium outcome. It has to be assumed that $\lambda\left(X-1\right)\frac{\sqrt{2}-1}{2} > \gamma$, in order to obtain an interior solution

Proposition 4 *Compared to the welfare maximizing amount of information acquisition, the outsider underinvests and the insider overinvests. The profits of both players are higher than in the welfare optimal equilibrium.*

Proof. Deriving equation 2.30 with respect to $q_1$and q_2 and solving the first order conditions yields the welfare maximizing amount of information acquisition:

$$q_1 = q_2 = q_{FB} = \frac{\lambda\left(X-1\right)}{\lambda\left(X-1\right)+2\gamma}. \tag{2.31}$$

Simple calculations result in:

$$q_1^* < q_{FB} < q_2^*. \tag{2.32}$$

Plugging the equilibrium effort q_1^* and q_2^* from 2.26 and 2.27 in the profit function 2.24 gives the equilibrium profits:

$$\Pi_2\left(q_2^*, q_1^*\right) = \frac{\left(X-1\right)^2\lambda^2\left(2\gamma+\left(X-1\right)\lambda\right)^2}{16\gamma\left(\gamma+\left(X-1\right)\lambda\right)^2}, \tag{2.33}$$

$$\Pi_1\left(q_1^*, q_2^*\right) = \frac{\left(X-1\right)^2\lambda^2}{4\left(\gamma+\left(X-1\right)\lambda\right)}. \tag{2.34}$$

The profits for welfare maximizing information acquisition are

$$\Pi_i\left(q_{FB}, q_{FB}\right) = \frac{\gamma\left(X-1\right)^2\lambda^2}{\left(2\gamma+\left(X-1\right)\lambda\right)^2}. \tag{2.35}$$

Simple algebraic manipulations prove that:

$$\Pi_2\left(q_2^*, q_1^*\right) > \Pi_1\left(q_1^*, q_2^*\right) > \Pi_i\left(q_{FB}, q_{FB}\right). \tag{2.36}$$

■

If welfare is lower and the banks' profits higher in the asymmetric equilibrium, an immediate consequence of proposition 4 is that the expected interest rate the firm has to pay will be higher.

2.3.3 Discussion

The model explains why relationship banking is observed in countries with a few closely related banks. Strategic information acquisition will lead to asymmetric situations, where only one bank specializes on a firm. This bank acquires a lot of information about the firm. However, the firm will not benefit from this information acquisition. The informational advantage enables the housebank to discourage competition and ask for high interest rates. Welfare will be reduced. The only way for the firms to prevent this inefficiency is to keep the potential lenders anonymous, or at least to inhibit the exchange of verifiable information between them. In a narrow banking market, this will not be possible.

Note that the effect identified with this model is more general. In almost all common value auctions, the bidders have the possibility to inform themselves more precisely about the true value of the object. A classical example has been the selling of oil drilling rights. Similarly to the banking case, the specialization of bidders may decrease the seller's profit.

2.4 Symmetric equilibrium with n bidders

In this section, it will be assumed that the bidders do not observe the quality of each other's information. Therefore, information acquisition and bidding will be modeled to take place simultaneously. It is shown that a symmetric equilibrium exists in this case. Again, the symmetric equilibrium without information acquisition will be calculated first. All n bidders are assumed to receive information with the same probability q and to bid with identical bidding parameters $F(b), H(b), \mu$. The indifference conditions are now for the case of a good signal:

$$\begin{aligned}
\pi^g(b) &= (b-1)\left[\sum_{s=0}^{n-1}\binom{n-1}{s} q^s (1-F(b))^s (1-q)^{n-1-s}\right. \qquad (2.37)\\
&\quad \cdot \left.\left[\sum_{t=0}^{n-1-s}\binom{n-1-s}{t}(1-\mu)^{n-1-s-t}\mu^t(1-H(b))^t\right]\right]\\
&= (b-1)\left[\sum_{s=0}^{n-1}\binom{n-1}{s} q^s (1-F(b))^s (1-q)^{n-1-s}\left[1-\mu H(b)\right]^{n-1-s}\right]\\
&= (b-1)\left[q(1-F(b))+(1-q)\left[(1-\mu)+\mu(1-H(b))\right]\right]^{n-1}\\
&= (b-1)\left[1-qF(b)-(1-q)\mu H(b)\right]^{n-1}.
\end{aligned}$$

In case the bank has not received any information, its profit is:

$$\begin{aligned}
\pi^{0}(b) &= \lambda \sum_{s=0}^{n-1} \binom{n-1}{s} q^{s} (1-F(b))^{s} (1-q)^{n-1-s} \left[1-\mu H(b)\right]^{n-1-s} (b-1) \\
&\quad - (1-\lambda) \sum_{s=0}^{n-1} \binom{n-1}{s} q^{s} (1-q)^{n-1-s} \left[1-\mu H(b)\right]^{n-1-s} \\
&= \lambda (b-1) \left[q(1-F(b)) + (1-q)\left[1-\mu H(b)\right]\right]^{n-1} && (2.38) \\
&\quad - (1-\lambda) \left[q^{s} + (1-q)\left[1-\mu H(b)\right]\right]^{n-1}. && (2.39)
\end{aligned}$$

Some preliminary definitions will again facilitate the exposition of the equilibrium.

Definition 2

$$\hat{b}_n := \frac{1-\lambda}{\lambda (1-q)^{n-1}} + 1, \tag{2.40}$$

$$\hat{\mu}_n := \left[1 - \frac{q}{1-q} \frac{(1-\lambda)^{\frac{1}{n-1}}}{(\lambda (X-1))^{\frac{1}{n-1}} - (1-\lambda)^{\frac{1}{n-1}}}\right]. \tag{2.41}$$

The equilibrium bidding strategies are summarized in the following proposition

Proposition 5 *The equilibrium strategies are*
a) If $X \leq \hat{b}_n$, banks bid only after having received a good signal:

$$F(b) = \frac{1}{q}\left[1 - (1-q)\left(\frac{X-1}{b-1}\right)^{\frac{1}{n-1}}\right]. \tag{2.42}$$

b) In the case $X > \hat{b}_n$, banks bid with good and with probability $\hat{\mu}_n$ with inconclusive signals using the distribution functions

$$F(b) = \begin{cases} 0 & \textit{for } b < \frac{1}{\lambda}, \\ \frac{1}{q}\left[1 - \left(\frac{1-\lambda}{\lambda (b-1)}\right)^{\frac{1}{n-1}}\right] & \textit{for } b \in \left[\frac{1}{\lambda}, \hat{b}_n\right], \\ 1 & \textit{for } b > \hat{b}_n, \end{cases} \tag{2.43}$$

$$H(b) = \begin{cases} 0 & \textit{for } b < \hat{b}_n, \\ \frac{1}{\hat{\mu}_n}\left[1 - \frac{q}{1-q} \frac{(1-\lambda)^{\frac{1}{n-1}}}{(\lambda (b-1))^{\frac{1}{n-1}} - (1-\lambda)^{\frac{1}{n-1}}}\right] & \textit{for } b \in \left[\hat{b}_n, X\right], \\ 1 & \textit{for } b > X. \end{cases} \tag{2.44}$$

Proof. See Appendix A ■

Again, the most relevant situation is a). The decision to bid without a good signal now depends not only on the quality of the firms, but also on the number of potential lenders. For higher n, the value of $\hat{b}_n$ converges very quickly to infinity, so that even for high λ and only a few bidders, normally nobody would bid without having received a good signal. The figures 2.4, 2.5 and 2.6 show the evolution of the distribution functions for $n = 2, 6, 9$ and $\lambda = 0.9$, $q = 0.3$, $X = 3$.

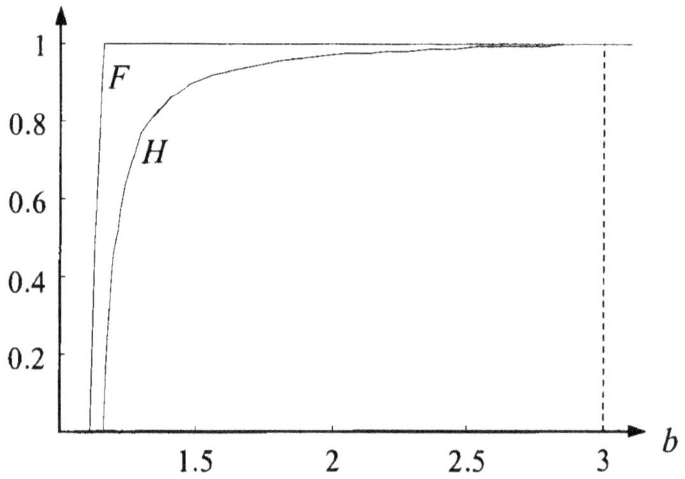

Figure 2.4: Symmetric equilibrium with 2 bidders, $\lambda = 0.9$, $q = 0.3$, $X = 3$.

Interestingly, the average bid made by a bidder increases with the number of competitors. Inserting the equilibrium strategies into the profit equation results in the following equilibrium profits:

Proposition 6 *The equilibrium profits of a bank are in case a)*

$$\pi = \lambda q (X - 1) (1 - q)^{n-1}, \tag{2.45}$$

and in case b)

$$\pi = q (1 - \lambda). \tag{2.46}$$

Proof. See Appendix A ■

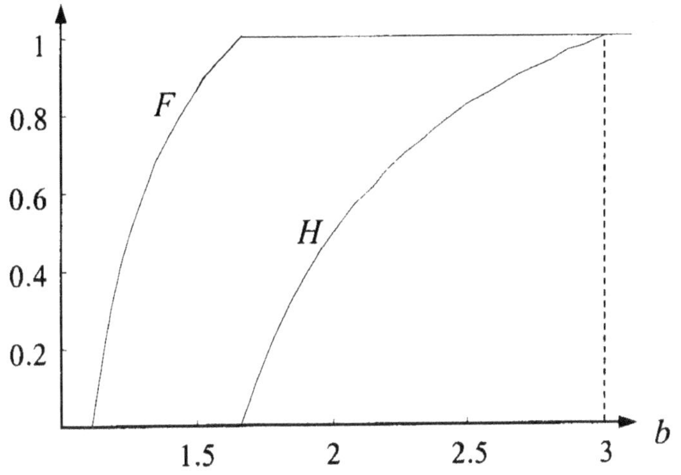

Figure 2.5: Symmetric equilibrium with 6 bidders, $\lambda = 0.9$, $q = 0.3$, $X = 3$.

Perhaps surprisingly, in case b) the profit of each bidder does not depend on the number of competitors. Of course, if there are more banks with symmetric strategies, each bank is receiving the credit with a lower probability. However, at the same time all banks are asking higher interest rates. Both effects exactly offset each other.

Without taking information acquisition into account, a higher number of banks increases the amount of information in the market and leads to stronger competition. In this case, welfare increases with the number of market participants.

Corollary 3 *Keeping the information of each bank fixed, welfare and the firm's profit increases with the number of banks. For $n > \ln\left(\frac{1}{1-q}\right)$, the profit of the banking sector decreases.*

Proof. If there are n banks with information q, the probability of a good project not to be financed is $(1-q)^n$. Hence the welfare produced is

$$W(q,n) = \lambda (X-1)\left[1-(1-q)^n\right], \tag{2.47}$$

which is increasing in n.

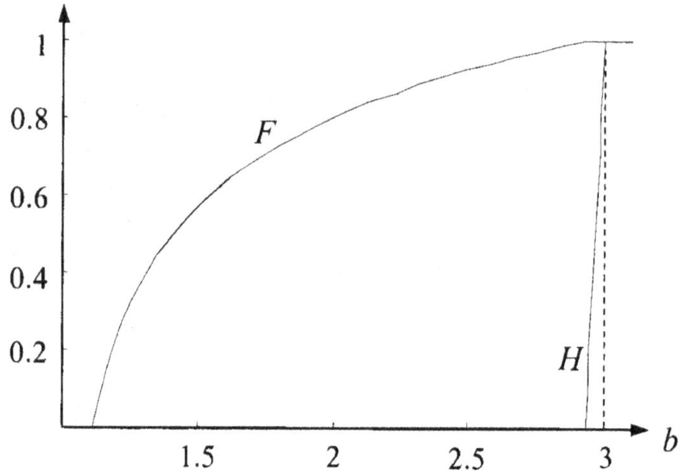

Figure 2.6: Symmetric equilibrium with 9 bidders, $\lambda = 0.9$, $q = 0.3$, $X = 3$.

The sum of the banks' profits is

$$n * \pi = n * \lambda q (X-1)(1-q)^{n-1}, \tag{2.48}$$

$$\frac{\partial}{\partial n} \lambda q (X-1)(1-q)^{n-1} * n \tag{2.49}$$

$$= \lambda (X-1)(1-q)^{n-1} q (1 + n \ln(1-q)) > 0, \tag{2.50}$$

$$\Leftrightarrow q < 1 - e^{-\frac{1}{n}} \tag{2.51}$$

Only if q and n are small an increase in the number of banks may increase the collective profit of the banking sector.

The firm's profit is changing with n as:

$$\frac{\partial}{\partial n} [W(q,n) - n * \pi] = -\lambda (X-1)(1-q)^{n-1} [q + (1 + q(n-1)) \ln(1-q)]$$

For $n \geq 1$, this is positive for all $q \in (0,1)$. ■

2.4.1 Simultaneous information acquisition

A strategy for bank j in the simultaneous choice game consists of (p_j, F_j, H_j, μ_j). In equilibrium, deviating from this strategy while keeping the other players strategies fixed, should not be profitable. However, the profit of bank j is not influenced by its choice of its own distribution functions and μ_j, as it is indifferent between any bids on the support and makes a lower profit out of the support. Once bank j has made the decision about its investment in information acquisition, its profit only depends on the competitors' actions. Therefore, in order to prove that a given strategy combination is an equilibrium, one only has to check that secretly deviating from the choice of q_j does not increase the bank's profit. If all other banks have invested q, its profit as a function of his own choice of q_i will be in case a)

$$\Pi(q_i, q) = \lambda q_i (X-1)(1-q)^{n-1} - \gamma q_i^2, \tag{2.52}$$

and in case b)

$$\Pi(q_i, q) = q_i (1-\lambda) - \gamma q_i^2. \tag{2.53}$$

Deriving with respect to q_i and imposing symmetry leads to the following conditions for an equilibrium.

$$\begin{aligned} \frac{\partial}{q_i}\Pi(q_i, q) &= \lambda (X-1)(1-q)^{n-1} - 2\gamma q_i = 0, & (2.54)\\ &\Rightarrow \lambda (X-1)(1-q)^{n-1} - 2\gamma q = 0, & (2.55) \end{aligned}$$

for case a) and

$$\frac{\partial}{q_i}\Pi(q_i, q) = 1 - \lambda - 2\gamma q_i = 0. \tag{2.56}$$

in case b). Case b) can be solved trivially. The equilibrium acquisition of information is $q_i = \frac{(1-\lambda)}{2\gamma}$. Since the profit does not depend on the number of bidders, this does not depend on the number of bidders either.

However, in case a) an analytic solution of 2.55 is not possible. Nevertheless, some welfare results can be derived without explicitly solving the first order condition.

2.4.2 Welfare and Consumer Surplus

In case a), no bank bids without having received a signal. If there are n banks, screening with intensity q, the probability of a good project not to be financed is then $(1-q)^n$. Hence the welfare produced is

$$W(q, n) = \lambda (X-1)\left[1 - (1-q)^n\right] - n\gamma q^2. \tag{2.57}$$

The welfare optimal choice of q is solution of

$$\begin{aligned} \frac{\partial}{\partial q} W(q, n) &= \lambda (X-1)(1-q)^{n-1} n - 2n\gamma q = 0, & (2.58)\\ &\Leftrightarrow \lambda (X-1)(1-q)^{n-1} - 2\gamma q = 0. & (2.59) \end{aligned}$$

The first order condition 2.59 turns out to be identical to condition 2.55, defining the equilibrium intensity of information acquisition. This proves that:

Proposition 7 *In case a) the banks choose the welfare maximizing intensity of information acquisition in the equilibrium.*

The firm's average profit can be obtained by subtracting the banks' profit, which is n times the profit of one bank, given in equation 2.52, from the total welfare:

$$CS(q,n) = \lambda(X-1)\left[1-(1-q)^n\right] - n\lambda q(X-1)(1-q)^{n-1}. \tag{2.60}$$

Case b) is more complicated, as here the firms may be financed even if no bank gets a signal. The welfare produced is now

$$\begin{aligned} W(q,n) &= \lambda(X-1)\left[1-(1-q)^n(1-\hat{\mu}_n)^n\right] && (2.61)\\ &\quad -(1-\lambda)\left[1-(1-(1-q)\hat{\mu}_n)^n\right] - n\gamma q^2. && (2.62)\end{aligned}$$

The firm's profit can again be calculated by subtracting the profit of the banking sector from the overall welfare.

A numerical example shows that both, welfare and consumer surplus, do not necessarily increase with the number of banks. In fact, they seem to increase in case a) and to decrease in case b). Figure 2.7 depicts the welfare and consumer surplus for $X=2, \gamma=2$ and $\lambda=0.4$. This is an ex ante unprofitable sample, therefore only case b) applies. Welfare and consumer surplus increase with the number of competitors.However, this is not true for the case b). In figure 2.8, the fraction of good projects is $\lambda=0.8$, hence this is a profitable sample. For $n \leq 28$, banks sometimes bid without having received a signal. In this case welfare and consumer surplus decrease with the number of banks.

The fact that a higher number of competitors may decrease social welfare is somewhat counterintuitive. The reason is of course that information acquisition is dissipative and duplication of research effort therefore inefficient. However, the following proposition shows that eventually welfare will increase again with the number of banks and exceed the initial levels.

Proposition 8 *For $n \to \infty$, welfare converges to $\lambda(X-1)$.*

Proof. It has been shown in proposition 5 that for a sufficiently high number of banks only case a) applies. Define $\alpha < 1$ as the probability that the market accepts a good project. Since $\alpha = 1-(1-p)^n$, the total cost of achieving α with n banks is $n\gamma\left[1-\sqrt[n]{1-\alpha}\right]^2$, which converges to zero for $n \to \infty$ with fixed $\alpha < 1$, and to one for fixed n and $\alpha \to \infty$.

The total surplus generated is $\alpha\lambda(X-1) - n\gamma\left[1-\sqrt[n]{1-\alpha}\right]^2$. Assume that for all n the optimal welfare does not exceed a given level $W < \lambda(X-1)$.

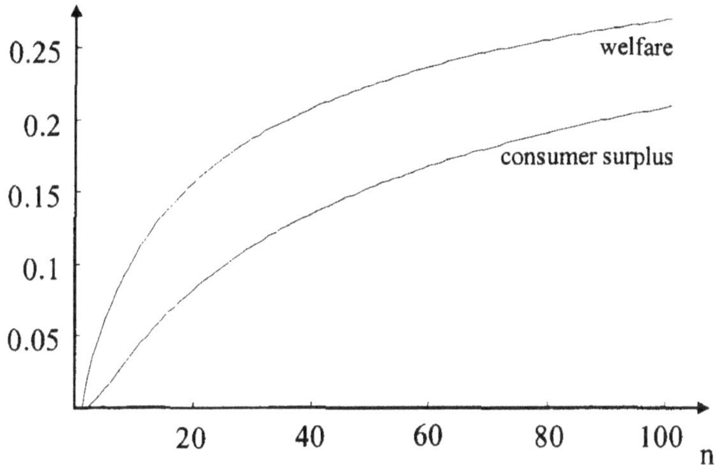

Figure 2.7: Case a) Welfare and Consumer Surplus for $\lambda = 0.4, \gamma = 2$ and $X = 2$

However, defining $\alpha = \frac{W+\lambda(X-1)}{2\lambda(X-1)} < 1$ and a n_α, such that $n_\alpha \gamma \left[1 - \sqrt[n_\alpha]{1-\alpha}\right]^2 < \frac{1}{2}\left[\lambda(X-1) - W\right]$ can be found.

It follows that $n_\alpha \lambda (X-1) - n_\alpha \gamma \left[1 - \sqrt[n_\alpha]{1-\alpha}\right]^2 > W$. Therefore the optimal welfare level for n_α must be higher than W. As this is true for every $W < \lambda(X-1)$, the optimal welfare level and therefore the level obtained by the market must converge to $\lambda(X-1)$ for $n \to \infty$. ■

If either the signals were not independent or if the cost for acquiring the first information unit became nonzero, this result would probably disappear. Wolfstetter and Elberfeld (1999) have analyzed a similar setting with fixed participation costs. They show that in the symmetric equilibrium bidders only decide to be active with a certain probability. In this situation welfare decreases with an increasing number of potential bidders.

2.4.3 Discussion

In the situations a) and b) welfare reacts very differently with respect to changes in the number of banks. In case a) the numerical simulations in figure 2.7 seem to indicate that welfare will in general increase with more banks. This is the situation where, without information acquisition, lending is not profitable. One

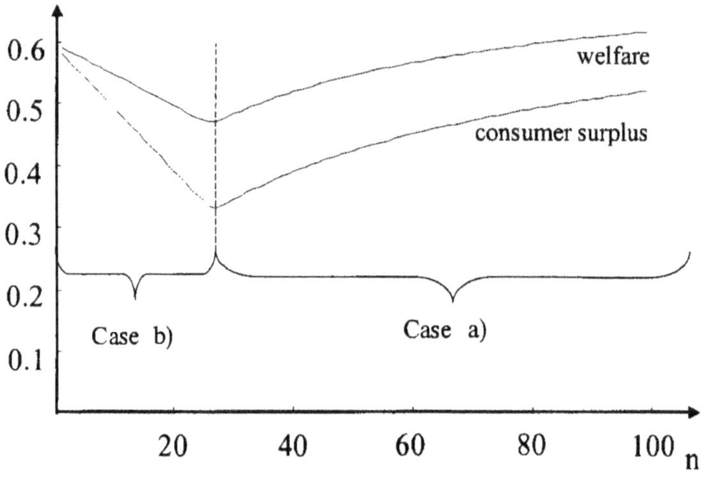

Figure 2.8: Case a) and b) Welfare and Surplus for $\lambda = 0.8, \gamma = 2$ and $X = 2$.

could interpret this as a sample of innovative firms. Initially the banks have very bad information about these firms. They know that some of the projects are worthwhile being financed, but they also know that there are a lot of lenders with unreasonable projects, most of which will fail. Without acquiring detailed information about the firms the banks will lose money. With a larger number of banks, every bank will reduce its spending on information acquisition. Nevertheless, in the aggregate, the probability rises that one of them will recognize the potential of the project and provide finance.

In case b), the fraction of good firms in the sample is so high that lending is profitable even without additional information. As the numerical simulation in figure 2.8 demonstrates, welfare will then in general decrease with the number of banks. This is because in a situation where lending is possible without additional information, the acquisition of information is less socially valuable. The main reason for banks to collect information is not to be able to finance firms, but rather to get a competitive advantage. An additional bank will increase the overall expenditures on information acquisition without substantially increasing the probability of a firm getting a credit. Only for very large numbers, the banks will not bid any more without having received information and welfare will rise again. This scenario could be interpreted as lending to mature firms in a stable environment. Under these circumstances, banks have sufficient prior information in order to select a pool of ex ante profitable firms. Two banks seem then to be sufficient to ensure an information efficient and reasonably competitive financial market.

These results seem to suggest that the optimal structure of the banking sector depends on the economy. A given banking sector performs differently for different groups of firms. This is visualized on figure 2.9. The fraction of the maximal surplus $\lambda(X-1)$ realized by the competitive equilibrium has been plotted as a function of the quality of the sample λ. A dispersed banking sector generates more surplus with innovative firms i.e. for low λ, whereas a more concentrated banking sector mainly lends to conservative industries.

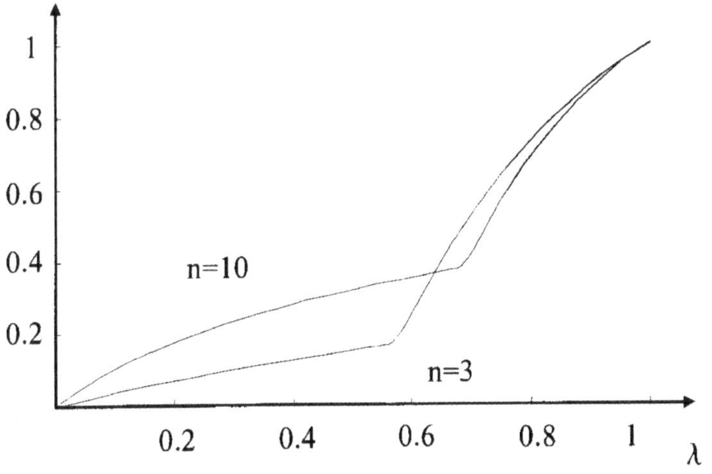

Figure 2.9: Surplus over $\lambda(X-1)$ as a function of λ for $X=2, \gamma=2$.

Again, the auction theoretic framework can be applied to more general contexts. For example, the model could explain why initial public offerings are often allocated by investment banks among a rather small number of preferred investors.

2.5 Common and private values

So far, credit market competition has been modeled as a pure common value auction. The main motivation for the common value approach is the credit risk. In addition, there may be other common value components, such as the repayment date, expected fee income related to the credit or the probability of giving another credit.

However, credit market competition may also contain private value aspects. A difference in private values means that two banks are able to generate different revenues with the credits of the same firm. Again this can happen for several reasons. For example, banks might have sector or firm specific knowledge that enables them to detect and recall risky credits earlier or to carry out a more efficient bankruptcy management. This possibility is analyzed in detail in chapter 5. However, the most straightforward reason for differences in private values is that banks have different refinancing or opportunity costs for the money they lend. A bank with access to cheap funds or high excess liquidity makes a higher profit with the credit of a given firm than a bank that has high refinancing costs.

In the next chapter, open bid auctions will be considered. It will be demonstrated that differences in private values or refinancing costs have a crucial influence when bids are observable. In this section, it will be shown that this is not true for the sealed bid setting. The results obtained with sealed bids are robust with respect to the introduction of private value components. In addition, it will be demonstrated how private values can give an intuitive motivation for the mixed strategy equilibria of the auctions with discrete values. Similar to the Harsanyi (1973) purification method, the mixed equilibria can be approached by pure strategy equilibria in a model with additional private value components.

2.5.1 The model

In principle the same setting as above is used, except that now the banks' refinancing cost v are assumed to be independently uniformly distributed on $[1, 1+r]$. It is not unreasonable to assume that the banks' refinancing costs lie in a well defined interval. A bank's depositors will probably not lend money for negative interest rates, which determines the lower bound. The maximal interest rate r could for example be the interbank rate.

Only the simple case of a symmetric duopoly is analyzed. The two banks participating in the auction receive information with the same probability p. The restriction $\lambda X < 1$ is imposed, so that lending without additional information is not profitable.

Proposition 9 *In equilibrium a bidder with a refinancing cost of v places the bid:*

$$b_r(v) = \frac{2(1-q)Xr - q(r-v-1)(1+r+v)}{2q(v-1)-2r}.$$

Proof. See Appendix A ■

Proposition 10 *For $r \to 0$, the bid distribution converges against the mixed strategy equilibrium distribution for the pure common value auction, i.e.*

$$\lim_{r \to 0} \left[\Pr\left(b_r\left(v\right) < b\right)\right] = \frac{1}{q}\left(1 - \frac{\left(1-q\right)\left(X-1\right)}{b-1}\right).$$

Proof. See Appendix A ■

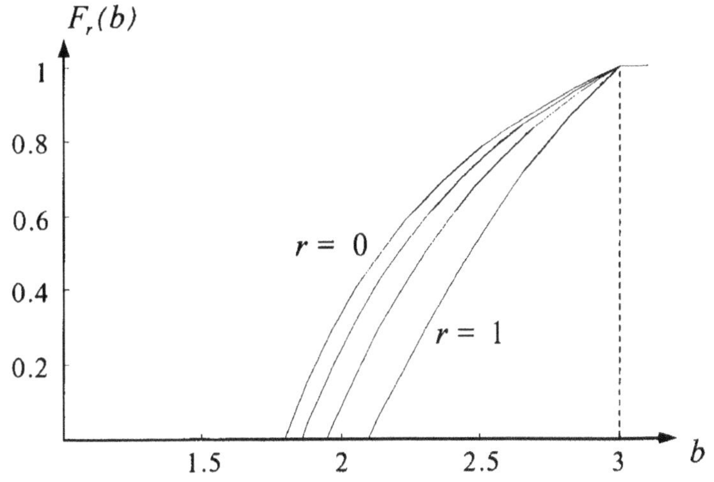

Figure 2.10: Bid distributions for $r = 1$, $r = 0.5$, $r = 0.2$ and $r = 0$.

Figure 2.10 shows the bid distributions resulting from different variances of the refinancing costs. With a larger variance of the refinancing costs the distribution of the bids becomes steeper. Therefore, the bidders are not indifferent any more, but there is an optimal bid for every possible refinancing cost. The bidders with the low refinancing costs are especially eager to give the credit. Therefore, they will bid at the lower end of the bidding support, where the probability of the competitor bidding in this region has become smaller. Bidders with higher refinancing costs evaluate differently the trade-off between getting the credit and asking a high margin. They will only want to give the credit if they obtain a margin substantially above their refinancing costs. Thus, there is an optimal bid for every value of the refinancing costs. Since the refinancing costs are randomly distributed, this results in a random distribution of the bids. In the limit, if the support of the banks refinancing costs becomes smaller, this distribution converges to the bid distribution in the pure common value auction.

Extending this result to the case of more general distributions is a challenging task for future work. Probably the bid distributions will converge to the common value distribution for any distribution of private values the support of which converges to zero.

2.6 Conclusion

The model has shown that taking into account the banks' incentives for screening firms helps to understand a range of phenomena. Most importantly, it has been demonstrated that good relationships between banks and good information of the banks about each other will lead to a specialization of one bank on a firm. This bank will acquire a lot of information about the firm and establish a housebank relationship. The other banks then have no incentives to increase competition by also acquiring information about the firm. This would not happen if the banks did not know the competitors' effort to acquire information. In this case the first best information acquisition effort would be invested. Hence competition in banking can be reduced quite easily. In order to partition the market no tacit collusion with punishment threads is necessary, club like relationships between banks are sufficient. Once established, tight housebank relationships can also serve as a signal that this bank is already an insider for a given firm and prevent other banks from acquiring information about this firm. The board membership of a banker may be an especially clear signal.

The second result shows that increasing the number of competitors may not always be the best thing for improving competition and welfare. If there are a lot of banks, a lot of money will be spent on information acquisition. This will not be welfare increasing if the information produced is socially not very valuable, because the sample of firms is so good that it could be financed without additional information.

However, more banks and more information acquisition may be beneficial for the innovative sectors of an economy. Here information acquisition is much more important for producing welfare. A larger number of banks will increase the chance of one bank detecting the value of an innovative project. Therefore regulations such as the Interstate Banking Act, which is responsible for an artificially large number of banks in the USA, may not be necessarily welfare increasing, but will boost the innovative sector of the economy.

Chapter 3

Open bid auctions

3.1 Introduction

So far, all existing auction models of bank competition have used a sealed bid framework. However, the assumption that bank competition is a sealed bid auction implies that banks are only allowed to make one offer and do not receive any information about the rival's bid. For large firms this is not very realistic. In reality, the interest rate is determined in a process involving repeated bidding and information revelation. The precise features of this bidding process depend very much on the specific case. However, there are several reasons to assume that in general this process is rather an open than a sealed bidding. Banks certainly have incentives to learn the offer of another bank and seem sometimes to be able to do that. The firms are interested in open bidding as well. It seems to be common that firms use the credit offer of one bank to incite another bank to undercut. They can credibly communicate the offer for example by showing the credit contract. In fact, they do not even need to do so. We also obtain an open bid setting when a bank, realizing that its credit offer will not be accepted, makes a second offer. Of course in order to obtain the perfect analogy of an open bid auction, firms would have to go back and forth from one bank to the other infinitely often, which is definitely not realistic. In fact, firms also switch part of their bank loans quite frequently from one bank to another, depending on the offers and on the outcomes of the negotiations with the banks they do business with. In the case of small firms, this may be less frequent than in the case of larger firms. The case of switching infinitely often is thus a limiting case.

The truth probably lies somewhere between pure open and sealed bidding. The appropriate model would be auctions with costly bidding as modeled by Hirshleifer and Kent (1996) or auctions with a limited number of bids as in Kamecke (1997). It is argued in this chapter that, at least for big firms, the flavor of the process determining the interest rate is better captured in the open bid setting. Qualitative results should carry over to the more realistic cases.

Whether or not one bank can observe the interest rate offered by the other bank makes a big difference for the profits of the players as well as for the ex ante incentives of information acquisition. The observation of the insider's bids gives the outsider information about the firm's value. This should reduce the insiders advantage and increase the outsider's profit. Fisher (1990) has concluded that informational rents should disappear. However, the model in this chapter shows that, at least for a low number of banks, informational rents can be obtained despite open bids.

However, the situation drastically changes if a sufficient number of outside banks is introduced. Whereas in the sealed bid setting a better informed bank has always higher profits, it will be shown in this chapter that in an open bid setting, better information can decrease profits. In markets with a large number of outsiders, the informational advantage turns out to be a disadvantage. The bank would prefer to forget its information or reveal it to the competitors. This may explain why in very fragmented markets, such as in the United States, less information intensive short-term commercial credit is predominant.

In the first chapter, the mathematical difficulties of asymmetric auctions have been reduced by the discrete information acquisition technology which only allows for perfectly informed or uninformed bidders. In this chapter, the mathematical structure has been further simplified. It is assumed that one bidder has more information than the others. A similar setting in which one perfectly informed bidder competes with several less informed outsiders in a sealed bid common value auction has been recently analyzed by Kagel and Levin [60]. In the open bid setting of this chapter, the insider's strategy is simple to derive. Because the insider is not able to learn anything from the bidding behavior of the outsider, his dominant strategy is to leave the auction when the price equals his own valuation. However, in a pure common values auction with open bidding and perfectly asymmetric information the outsider's strategy cannot be determined.

In fact, the outsider is indifferent between all possible strategies. He will not make profit for any dropout prices because he either doesn't get the credit or gets it exactly for the break-even price. Without determining the outsider's strategy, it is impossible to calculate the insider's profit.

This Bertrand-like situation can be avoided with the introduction of private values. Then, the insider's offer will only be a noisy signal of the common value and one of the bidders will drop out earlier. By letting the private value component of our model tend to zero, we can obtain a result for the pure common value game as the limit of a disturbed auction.

3.2 The Model

In order to simplify the intuition and the comparability with standard auction theory, it is assumed that the firm is selling a bond to the bank in an English auction. This means that the bidding process is formalized in the usual way by assuming that the auctioneer (i.e. the firm) continuously raises the price (i.e. lowers the interest rate) until one of the two bidders (banks) decides to leave the auction. The credit is then assigned to the other bank at the price at which the auction had stopped.

In contrast to the previous model, now there are not only two possible realizations for the common values. It is assumed that the common value components of the firm's credit (i.e. risk etc.) are summarized in a random variable $\hat{c}$ which is assumed to be uniformly distributed on the interval $[\underline{c}, \overline{c}]$. In particular, this implies that the project of the firm is ex ante profitable. Unlike in the previous chapter, credit risk can always be compensated with a higher interest rate. One of the banks knows the realization of $\hat{c}$, and will therefore be called in the following the inside bank.

The private values (i.e. refinancing costs etc.) are assumed to be for each bank independently distributed on the interval $[\underline{p}, \overline{p}]$. Each bank only knows its own private value.

In the main part of the chapter, only a duopoly with one insider bidding against one outsider is analyzed. The realization of the insider's private value will be denoted with p_{ins} and the outsider's private value with p_{out}. The profit of a bank having paid the price π is then $p_{ins} + \hat{c} - \pi$ and $p_{out} + \hat{c} - \pi$, respectively.

This model has equilibria in pure strategies. Depending on their knowledge, banks have to decide at which price they will drop out of the auction, given that the other bank has not left yet. Hence the strategy of the inside bank is a function $\pi_{ins} : [\underline{p}, \overline{p}] \times [\underline{c}, \overline{c}] \rightarrow R$ and the strategy of the outsider is a function $\pi_{out} : [\underline{p}, \overline{p}] \rightarrow R$.

Strategy of the inside bank Using the same arguments as in the standard second price ascending bid auction, it can be shown that the inside bank has a dominant strategy: As its own strategy does not influence the price it has to pay, it has no interest to stay in the auction after the price has risen above its valuation, but there is also no reason for leaving earlier. So the only dominant strategy for the insider is to leave the auction exactly at a price equal to its own valuation, i.e. at

$$\pi_{ins}(p_{ins}, \hat{c}) = p_{ins} + \hat{c}. \tag{3.1}$$

Strategy of the outside bank Observing that the insider is still in the auction at a price π, the outsider knows that $p_{ins} + \hat{c} > \pi$. This means that the common value and the private value of the insider are lying in the grey region of figure 3.1. However, the outsider is not interested in the expected value of $\hat{c}$ having observed π, but in the expected value of $\hat{c}$ conditional on winning the auction. He knows that, if he wins the auction, the above inequality must be an equality, i.e. that the common and private value of the insider lie on the line $\pi - p_{ins}$. He will leave the auction if the price equals the expected valuation under the condition that the insider leaves the auction at this price.

Setting $\Delta p := \overline{p} - \underline{p}$ and $\Delta c := \overline{c} - \underline{c}$ will simplify the notation in the remaining part of this chapter. When specifying the outsiders strategy, two cases with qualitatively different equilibria have to be distinguished: $\Delta p < \Delta c$ (figure 3.1) and $\Delta p \geq \Delta c$ (figure 3.3)

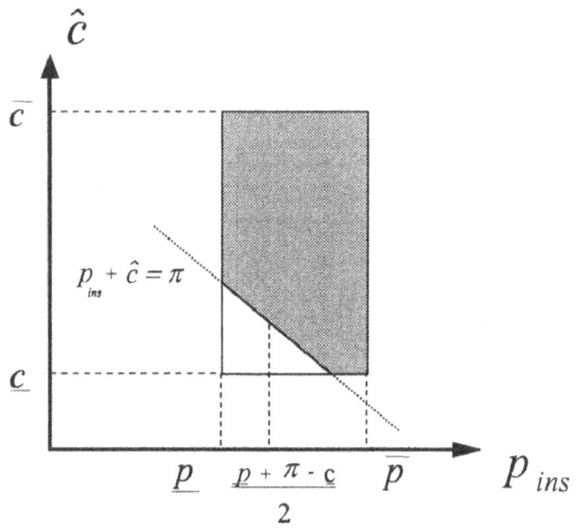

Figure 3.1: Case 1: $\Delta p < \Delta c$

Case 1: If $\Delta p < \Delta c$ and if the price is lower than $\overline{p} + \underline{c}$, the rectangle in figure 3.1 is cut in the lower left corner. The outsider knows that the private values of insider leaving the auction at $\pi < \underline{c} + \overline{p}$ are uniformly distributed on the interval $[\underline{p}, \pi - \underline{c}]$. As the insider drops out at the point where he makes zero profits, the outsider can make a positive expected profit if his private value is higher than the expected private value of the insider i.e. if $p_{out} > \frac{1}{2}\left(\underline{p} + \pi - \underline{c}\right)$. So the outsider will leave the auction when this inequality strictly holds i.e. his bidding strategy is $\pi_{out}(p_{out}) = 2p_{out} - \underline{p} + \underline{c}$.

If $\pi > \underline{c} + \overline{p}$ the situation changes. For $\pi \in \left[\underline{c} + \overline{p}, \overline{c} + \underline{p}\right]$ the price does not reveal new information about the private value of the insider. Thus, if the outsider has not dropped out before, he will stay in the auction until the insider leaves or the price rises above $\underline{p} + \overline{c}$.

For $\pi > \overline{c} + \underline{p}$, a similar reasoning as for $\pi < \underline{p} + \overline{c}$ applies. The rectangle of figure 3.1 is now cut in the upper right corner. The expected private value of the insider, conditional on the outsider winning the auction is $\frac{1}{2}(\overline{p} + \pi - \overline{c})$. Thus an outsider with $p_{out} < \frac{1}{2}(\overline{p} + \pi - \overline{c})$ will leave, his strategy therefore is $\pi_{out}(p_{out}) = 2p_{out} - \overline{p} + \overline{c}$.

Thus the outsider's strategy is

$$\pi_{out}(p_{out}) = \begin{cases} 2p_{out} - \underline{p} + \underline{c} & \text{for } p_{out} \leq \frac{1}{2}(\underline{p} + \overline{p}), \\ 2p_{out} - \overline{p} + \overline{c} & \text{for } p_{out} > \frac{1}{2}(\underline{p} + \overline{p}). \end{cases} \tag{3.2}$$

Figure 3.2 shows the bidding strategy of the outsider.

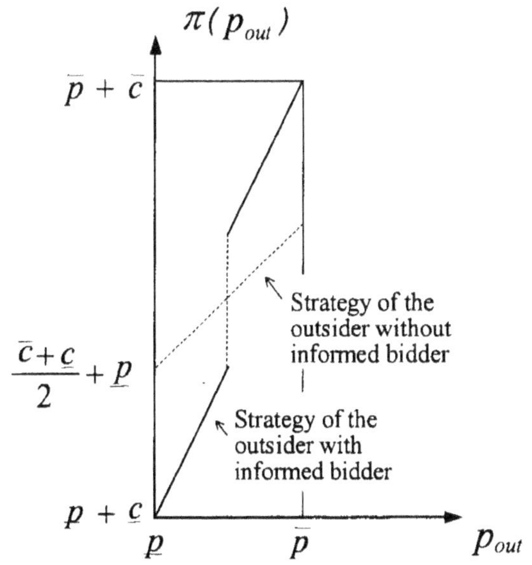

Figure 3.2: Bidding function of the outsider for $\Delta p < \Delta c$

Case 2: For $\Delta p \geq \Delta c$ the situation is represented in figure 3.3. Now, until the price reaches $\underline{p} + \overline{c}$ the rectangle is again cut in the left lower corner . This results in the same bidding function as in the previous case: $\pi_{out}(p_{out}) = 2p_{out} - \underline{p} + \underline{c}$.

For $\pi \in [\underline{p} + \overline{c}, \overline{p} + \underline{c}]$ however, the price line cuts the upper and lower side of the rectangle. Hence, the expected private value of a leaving insider is $\frac{1}{2}(\pi - \overline{c} + \pi - \underline{c})$. This means the outsider will drop out if $\pi - \frac{1}{2}(\overline{c} + \underline{c}) > p_{out}$ i.e. he will bid up to his ex ante expected valuation $\pi_{out}(p_{out}) = p_{out} + \frac{1}{2}(\overline{c} + \underline{c})$.

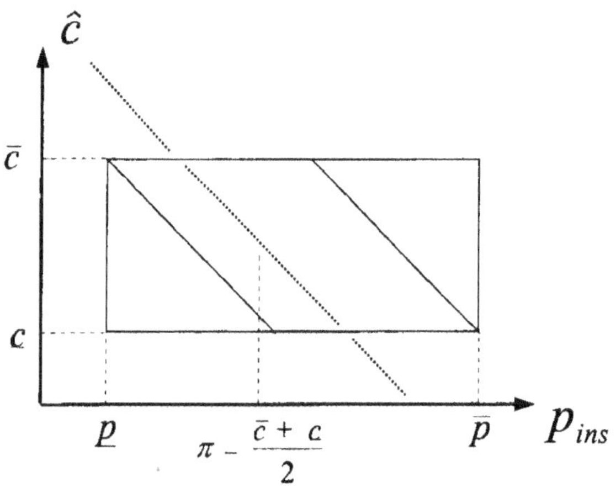

Figure 3.3: Case 2: $\Delta p \geq \Delta c$

If the price rises above $\overline{p} + \underline{c}$, situation is analogous to the previous section and the bidding function will be $\pi_{out}(p_{out}) = 2p_{out} - \overline{p} + \overline{c}$.

Summarizing this discussion, the bidding function has the following form:

$$\pi_{out}(p_{out}) = \begin{cases} 2p_{out} - \underline{p} + \underline{c} & \text{for } p_{out} < \underline{p} + \frac{1}{2}(\overline{c} - \underline{c}), \\ p_{out} + \frac{1}{2}(\overline{c} + \underline{c}) & \text{for } \underline{p} + \frac{1}{2}(\overline{c} - \underline{c}) \leq p_{out} < \overline{p} - \frac{1}{2}(\overline{c} - \underline{c}), \\ 2p_{out} - \overline{p} + \overline{c} & \text{for } p_{out} \geq \overline{p} - \frac{1}{2}(\overline{c} - \underline{c}). \end{cases} \tag{3.3}$$

This function is drawn in figure 3.4

Thus in general the outsider's strategy is neither differentiable nor continuous. Outsiders with low private values already drop out before the expected value $p_{out} + \frac{1}{2}(\overline{c} + \underline{c})$ is reached, i.e. they bid less aggressively than in the pure private value case. Outsiders with high private values stay in the auction even if the price rises above their ex ante expected valuation. Knowing that there is an informed insider, they behave much more aggressively.

3.2.1 The banks' expected equilibrium profits

By integrating over the different cases and strategies, the expected profits of the inside and the outside bank before learning its own private value can be

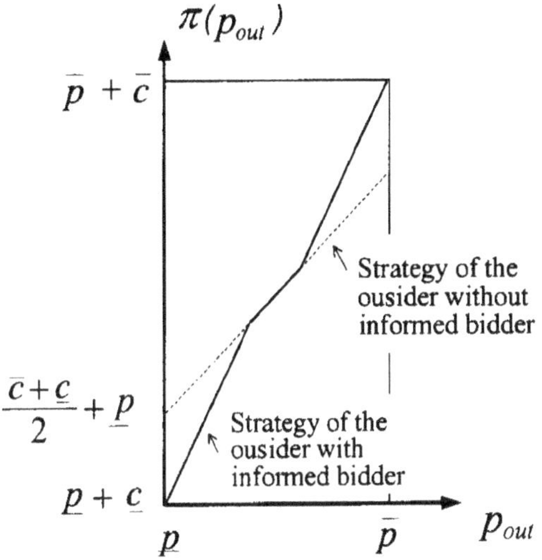

Figure 3.4: Bidding function of the outsider for $\Delta p > \Delta c$

calculated. This is not unreasonable, if the bank's private value is different for each firm, or quickly changes in time. The profits do not depend on the absolute levels, but only on the variance of common and private values Δp and Δc.

Proposition 11 *The expected profit of the inside bank is*

$$\Pi_{ins}(\Delta p, \Delta c) = \begin{cases} \dfrac{4\Delta p^3 + 4\Delta c^2 \Delta p - \Delta c^3}{24\Delta p^2} & \text{for } \Delta c \leq \Delta p, \\ \dfrac{6\Delta c^2 + \Delta p^2}{24\Delta c} & \text{for } \Delta c > \Delta p. \end{cases} \tag{3.4}$$

Proof. See Appendix B ■

Proposition 12 *The expected profit of the outsider is*

$$\Pi_{out}(\Delta p, \Delta c) = \begin{cases} \dfrac{8\Delta p^3 - 2\Delta c^2 \Delta p + \Delta c^3}{48\Delta p^2} & \text{for } \Delta c \leq \Delta p, \\ \dfrac{\Delta p\,(6\Delta c + \Delta p)}{48\Delta c} & \text{for } \Delta c > \Delta p. \end{cases} \tag{3.5}$$

Proof. See Appendix B ■

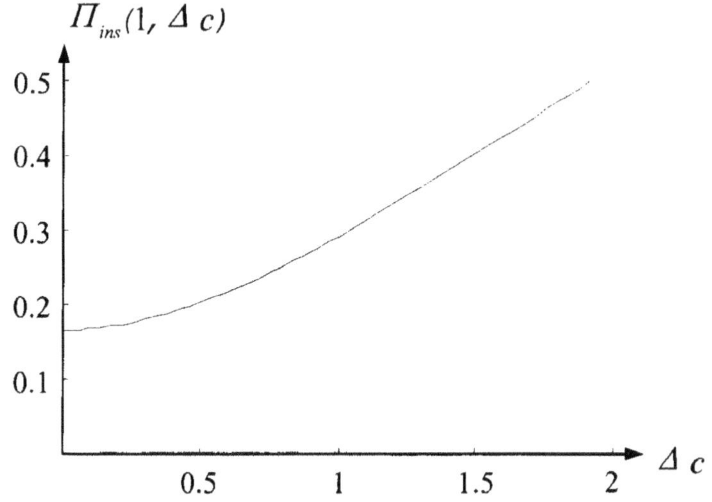

Figure 3.5: Profit of the inside bank as function of c

Fig 3.5 and. Fig 3.6 show the insider's and outsider's profit for $\Delta p = 1$, as a function of Δc. The results are consistent with standard auction theory. Setting $\Delta c = 0$, there are no common values and therefore no informational asymmetry. Both banks earn $\Pi_{ins}(\Delta p, 0) = \Pi_{out}(\Delta p, 0) = \frac{\Delta p}{6}$, which is the profit predicted for the independent private value auction with two bidders.

If the variances of private and common values are of about the same size i.e. $\Delta c \approx \Delta p$, the profits are $\Pi_{ins}(\Delta p, \Delta p) = \frac{7}{24}\Delta p$ and $\Pi_{out}(\Delta p, \Delta p) = \frac{7}{48}\Delta p$. The outsiders make approximately half the profit of the insiders.

For $\Delta p = 0$, the above solution concept does not apply any more. Without a private value component, the outsider would always want to leave at the same moment as the insider. However, the results in the mixed auction converge for $\Delta p \to 0$. For very small differences in private values, the profits are approximately $\Pi_{ins}(0, \Delta c) = \frac{1}{4}\Delta c$ and $\Pi_{out}(0, \Delta c) = 0$.

As expected, the insider's profit rises with the variance of the common value and the outsider's profit decreases with the informational advantage of the insider. However, it decreases less than the insider's profit increases. The implication of this result for the information acquisition of insider and outsider will be discussed in paragraph 3.2.3.

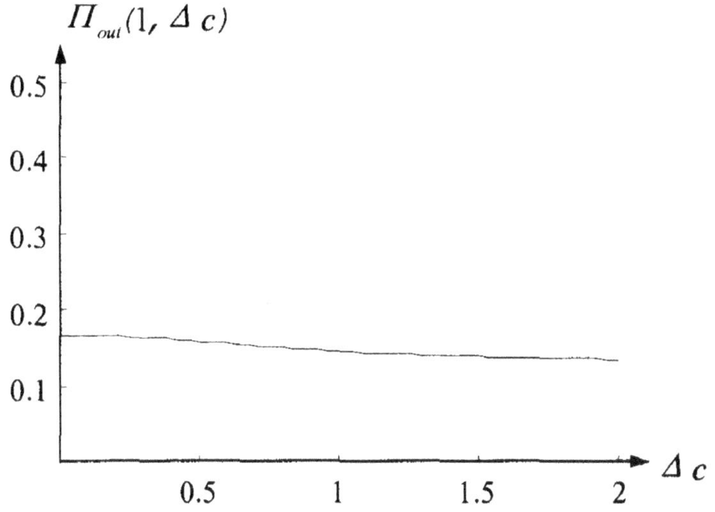

Figure 3.6: Profit of the outside bank as function of c

3.2.2 The firm's profit

If the auction were efficient, the firm's profit could be calculated as some constant, representing its expected profit, minus the bank's expected profits. This auction, however, is clearly not efficient as not always the buyer with the highest valuation gets the object. For high common values, outsiders with higher private values may drop out before insiders with lower private values, and for low common values the inverse may happen (which means that the outsider makes losses). It is easy to calculate the surplus that an efficient mechanism could create: The expected private value of the bank with the higher valuation equals $\underline{p} + \int_{\underline{p}}^{\overline{p}} x \cdot 2 \cdot \frac{x}{(\overline{p}-\underline{p})^2} = \underline{p} + \frac{2}{3}\left(\overline{p} - \underline{p}\right)$ and the expected common value is $\frac{1}{2}\left(\overline{c} + \underline{c}\right)$. As we only want to consider the effects of an increase in the variance of the private and common value component, we have to make sure that the expected value of the project and the expected evaluation of the firm have the same distance when we change Δp or c. We can obtain this by assuming that the firm's valuation is $\frac{1}{2}\left(\overline{p} + \underline{p} + \overline{c} + \underline{c}\right) - V$. Of course the model of this chapter only applies if the value of the credit to the bank is always higher than the firm's valuation of the credit i.e. for $\underline{p} + \underline{c} - V > 0$. Otherwise, the firm may not be willing to accept the credit at the conditions offered by the bank. The surplus that could be created by giving the object to the bank with the higher valuation is then $\underline{p} + \frac{2}{3}(\overline{p} - \underline{p}) + \frac{1}{2}(\overline{c} + \underline{c}) + V - \frac{1}{2}(\overline{p} + \underline{p} + \overline{c} + \underline{c}) = \frac{1}{6}\Delta p + V$.

The surplus created by the given auctioning mechanism can be calculated by adding the expected profit of the firm to the expected profits of the buyers. The firm's expected profit equals the expected price minus his valuation. The expected price can be calculated in a similar way as the banks' profits:

Proposition 13 *The expected price π is:*

$$\pi(\Delta p, \Delta c) = \begin{cases} \dfrac{\Delta c^3 - 4\Delta c^2 \Delta p + 12\Delta c \Delta p^2 + 8\Delta p^3}{24\Delta p^2} + \underline{p} + \underline{c} & \text{for } \Delta c \leq \Delta p, \\ \dfrac{1}{24}\left(6\Delta c + 12\Delta p - \dfrac{\Delta p^2}{c}\right) & \text{for } \Delta c > \Delta p. \end{cases} \tag{3.6}$$

Proof. See AppendixB ■

Corollary 4 *The firm's expected profit is:*

$$\Pi(\Delta p, \Delta c) = \begin{cases} V + \dfrac{\Delta c^3 - 4\Delta c^2 \Delta p - 4\Delta p^3}{24\Delta p^2} & \text{for } \Delta c \leq \Delta p, \\ V - \frac{1}{24}\left(6\Delta c + \dfrac{\Delta p^2}{\Delta c}\right) & \text{for } \Delta c > \Delta p. \end{cases} \tag{3.7}$$

A consistency check shows that for $\Delta c \to 0$, we obtain $\pi = \frac{\Delta p}{3}$, which is the expected price in a second price private value auction with two banks. For $\Delta p \to 0$, the expected price is $\pi = \frac{\Delta c}{4}$. As there are no more private values, there cannot be any inefficiencies. Therefore the surplus created must be first best i.e. V. From paragraph 3.2.1, it is known that only the insider makes a profit of $\frac{\Delta c}{4}$, so an amount of $V - \frac{\Delta c}{4}$ must be left to the firm. This is what we obtain with our formula for $\Delta p = 0$.

Summing up the profits of all participants, one obtains the created surplus

$$\Pi(\Delta p, \Delta c) + \Pi_{ins}(\Delta p, \Delta c) + \Pi_{out}(\Delta p, \Delta c) =$$

$$= \begin{cases} V + \frac{1}{48}\dfrac{\Delta c^3 - 2\Delta c^2 \Delta p + 8\Delta p^3}{\Delta p^2} & \text{for } \Delta c \leq \Delta p, \\ V + \frac{1}{48}\Delta p \dfrac{6\Delta c + \Delta p}{\Delta c} & \text{for } \Delta c > \Delta p. \end{cases} \tag{3.8}$$

Interestingly for $c \to \infty$, the efficiency loss quickly converges to $\left(\frac{1}{6} - \frac{1}{8}\right)\Delta p = 4.1667 \times 10^{-2}\Delta p$. The distribution of this surplus between the firm and the banks, however, changes rapidly in favor of the inside bank, as shown by figure 3.7. In this model, unlike in the first price setting of Campbell and Levin (1997), the seller never benefits from asymmetry. Indeed he has to pay most of the additional gains of the insider, whereas the outsider's profit is comparatively stable.

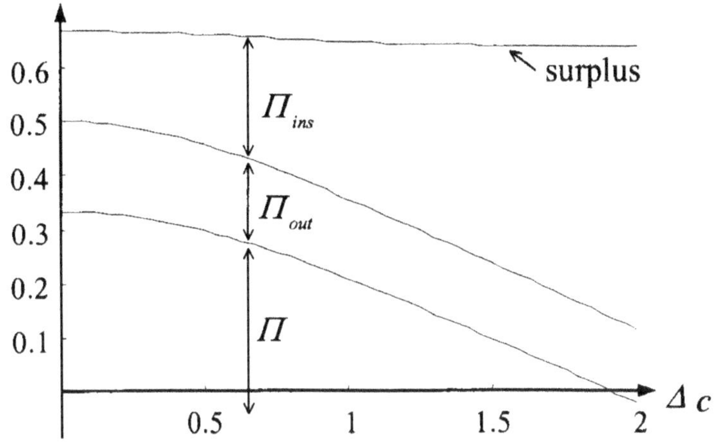

Figure 3.7: Total surplus and shares of participants for $p = 1$

3.2.3 Information acquisition

In order to evaluate the incentives for information acquisition of the insider and the outsider in a consistent way, it would be necessary to solve the model with more general than uniform distributions. Therefore, this section will only qualitatively discuss the implications of the model for information acquisition.

A situation where both banks have the same prior about the firm's true value is strategically identical to a situation in which both banks exactly know the firm's value. Both cases are pure private value auctions. If now one of the banks acquires more information about the firm, this bank will make a better estimation of the firms value. The expected value of the firm under the new information will shift in a direction unknown to the uninformed bank. The better is this information, the higher is the probability that the new estimate is far from the original one. From the point of view of the uninformed bank, the estimation of the better informed bank will be a random variable. The variance of this random variable will increase with the precision of the insider's information.

In the model, the possible range of the insider's estimate from the outsider' point of view is an interval of length Δc. Hence by acquiring information the insider should be capable to increase Δc. Leaving aside the costs of information acquisition, his marginal return from increasing his informational advantage is:

$$\frac{\partial}{\partial \Delta c}\Pi_{ins}(\Delta p, \Delta c) = \begin{cases} \dfrac{8\Delta c\Delta p - 3\Delta c^2}{24\Delta p^2} & \text{for } \Delta c \leq \Delta p, \\ \dfrac{6\Delta c^2 - \Delta p^2}{24\Delta c^2} & \text{for } \Delta c > \Delta p. \end{cases} \tag{3.9}$$

For fixed Δp this approaches asymptotically $\frac{1}{4}$, if Δc becomes very big. Even if the insider is already very well informed, additional information will benefit him. He can always improve his profit by acquiring more information.

The effects of the outsider's information acquisition are easier to understand. If the insider precisely knows the common value, whereas initially he only knows that it is uniformly distributed in an interval of the length c, then information acquisition should reduce the length of this interval.

$$-\frac{\partial}{\partial \Delta c}\Pi_{out}(\Delta p, c) = \begin{cases} \dfrac{4\Delta c\Delta p - 3\Delta c^2}{48\Delta p^2} & \text{for } \Delta c \leq \Delta p, \\ \dfrac{\Delta p^2}{48\Delta c^2} & \text{for } \Delta c > \Delta p. \end{cases} \tag{3.10}$$

His incentives to reduce the length of the interval are always smaller than the insiders incentives to increase it. For high Δc, his marginal return of information converges to zero. This means that the outsider's incentives to acquire information disappear if the insider is already very well informed.

3.2.4 Expected profit as a function of private values

So far, the expected profits of a bank before learning its own private value have been calculated. Nevertheless, it would be interesting to know how the banks' profits and their incentives for information acquisition change with their private values. The profit as function of the private value can easily be obtained by omitting one integration in the calculation of the expected profit. Due to the different cases, the mathematical functions are complicated. The graphical representations give a better idea of what is happening:

Whereas the insider's incentives to increase Δc slightly decrease with rising p_{ins}, the incentives of the outsider to decrease Δc are the biggest for intermediary p_{out}.

3.2.5 Several outsiders

Whereas the analysis with a general number of insiders and outsiders is difficult, the case of several identically informed outsiders bidding against one insider can be easily derived from the existing model.

The insider's strategy does not depend on the number of outsiders. The strategies of the outsiders are the same ones as in the two player case, as long

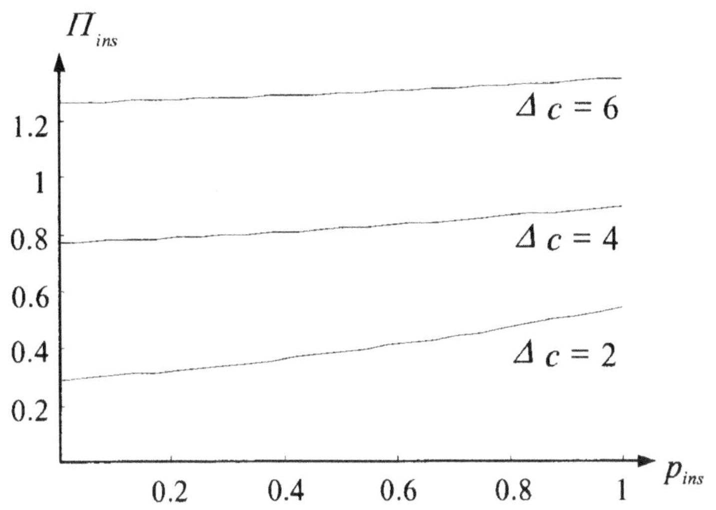

Figure 3.8: Insider's profit for $\Delta p = 1$ and $\Delta c = 2, 4, 6$

as the insider is still in the auction. However, they may continue bidding even if the insider has left the auction. Whereas it is difficult to calculate the expected profit of an outsider, the expected profit of the insider can be calculated as if he was bidding against one outsider whose private value is distributed according to the maximum value statistic. The resulting profit formulae are quite complex.

Proposition 14 *An insider bidding against n uninformed outsiders makes the following profit:*

$$\Pi_{ins}(\Delta p, \Delta c) =$$

$$= \begin{cases} \frac{1}{2^{n+2}\Delta p^{n+1}\Delta c(n+1)(n+2)(n+3)}\Big[2^{n+5}\Delta p^{n+3} \\ -\Delta c^{n+2}(\Delta c(1+n)^2 - 2(2+n)(3+n)\Delta p) - (2\Delta p - \Delta c)^{1+n} \\ (\Delta c^2(1+n)^2 + 16\Delta p^2 + 2\Delta c\Delta p(1+3n))\Big] & \text{for } \Delta c \leq \Delta p; \\ \frac{\Delta p^2}{\Delta c}\frac{8 - 2^{-n}(8+n(5+n))}{(n+1)(n+2)(n+3)} + \frac{\Delta c}{2^{n+1}} - \frac{\Delta p}{2^{n+1}}\frac{n-1}{n+1} & \text{for } \Delta c > \Delta p. \end{cases} \quad (3.11)$$

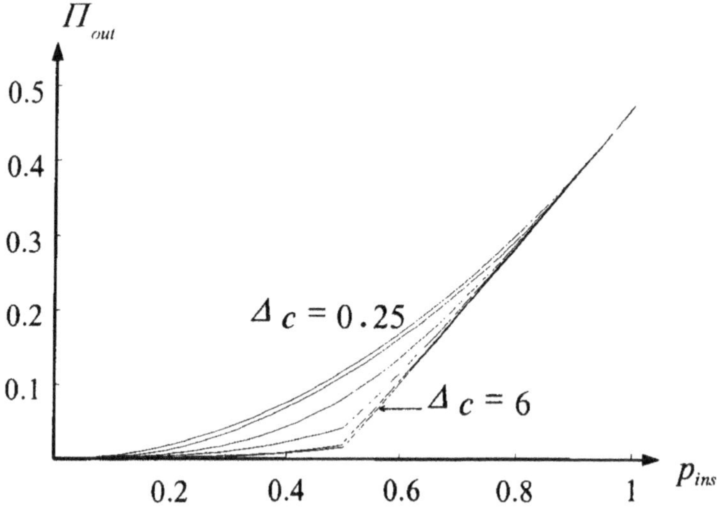

Figure 3.9: Outsider's profit for $\Delta p = 1$ and $\Delta c = 0.25, 0.5, 1, 2, 4, 6$

Proof. See Appendix B ■

For $\Delta c \to 0$, this converges correctly to the pure private value payoff $\frac{\Delta p}{(n+1)(n+2)}$. For $\Delta p \to 0$, the insider's profit is $\frac{\Delta c}{2^{n+1}}$. In the case of $\Delta p = \Delta c$, the profit formula can be simplified to $\Delta p \frac{2(4-2^{-n})}{(n+1)(n+2)(n+3)}$. The most interesting question is again how inside information influences the payoff. The simplest way to explore the properties of this complex function is by graphical representation.

Figures 3.10 and 3.11 show the profit as function of the informational difference between insider and outsider. The profit for identically informed banks has been plotted in the same graphic. Surprisingly the insider's profit falls below the profit he would make if all banks had the same information. The informational advantage turns out to be a disadvantage. The insider would either not want to acquire information, or if he had already acquired it, reveal it to all competitors. Only for very big Δc, i.e. for very big informational differences, information becomes advantageous again.

This is only surprising at the first view. It has been pointed out in section 3.2 that banks with high private values bid more aggressively than in the pure private value case. Hence, the higher the number of outsiders, the higher is the probability of an aggressive bank which can drive the insiders profit below his profit in the private value case.

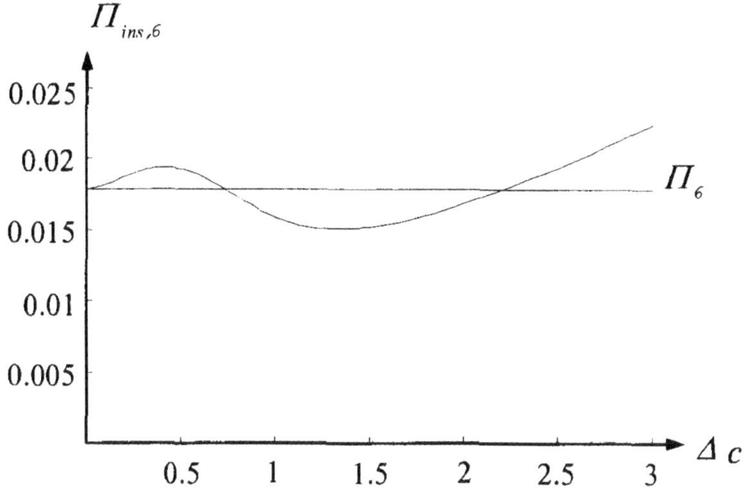

Figure 3.10: Profit of the insider with $n = 6$, $\Delta p = 1$.

Imagine for example that the private value component is very small compared with the variance of the common values which is the most relevant constellation for a credit auction. Then the outsiders which have a private value lower than the expected private value will leave the auction almost immediately, whereas all outsiders with a private value higher than the expected private value will stay almost until the price has reached the highest possible level. If there are a lot of banks the probability of at least one bidder having a private value higher than the expected private value is quite high, and hence the probability of the insider getting the object is quite low.

In fact, it can be shown that, if the number of banks is sufficiently high, every informational advantage turns into a disadvantage.

Corollary 5 *For every Δc, $\Delta p > 0$ there is an $\bar{n}$, such that for more than $\bar{n}$ banks the insider's profit is lower than his profit in the pure private value auction i.e. for $\Delta c = 0$.*

Proof. See Appendix B ■

Figure 3.12 shows the ratio of the insider's profit over the profit he would make as an uninformed bank for $\Delta p = 1$, $\Delta c = 2$. The quotient quickly falls below 1, implying that inside information becomes a disadvantage. This means

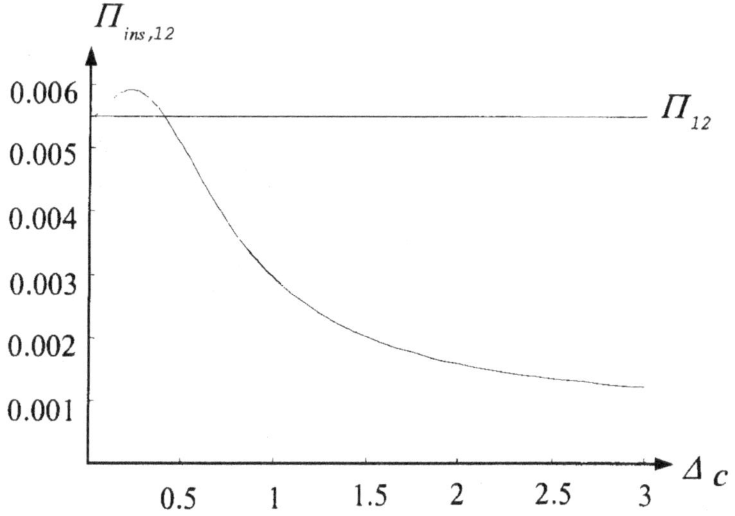

Figure 3.11: Insider's profit for $n = 12$, $\Delta p = 1$.

that in very crowded markets there is an equilibrium where no banks acquire information about a firm. Note that this is not necessarily the only equilibrium. Given that one bank is insider, other banks may have incentives to become insider as well.

3.3 Conclusion

This chapter has presented a conceptually simple model of the competition between banks with different information. In contrast to much of the literature, it has been assumed that banks learn the rival's offer.

It has been shown that, for high enough informational advantages or a low enough number of banks, informational rents survive even in an open bid model of banking competition. However, it has also been demonstrated that, with a high number of banks, their incentives for information acquisition may be very low. Indeed, they may become negative, since better information can even lower a bank's profit.

The model is too complex for endogenizing the information acquisition. Qualitative results seem to indicate that similar to the sealed bid case, with a low number of bidders, the incentives for information acquisition for the insider are

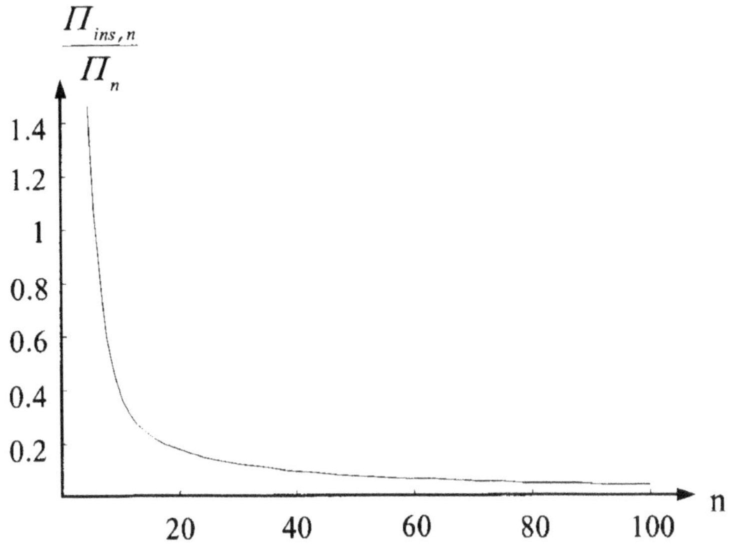

Figure 3.12: Insider's profit over profit in the symmetric auction. $\Delta p = 1$, $\Delta c = 2$.

bigger than for an outsider. Hence, informational asymmetry should arise endogenously with a low number of bidders, whereas a high number of bidders will tend to be symmetrically informed.

In the model, social welfare and the profit left to the firms is increasing with the number of banks and with the informational symmetry between them, but of course the welfare analysis is not complete. it has been implicitly assumed that information acquisition is socially wasteful as every project is worthwhile being financed. However, a typical problem for banks is that some projects might not be profitable at any interest rate. The existence of unprofitable projects should change the incentives for information acquisition as well as the welfare analysis. More research in this direction is needed.

How robust are the results? The assumption of uniform distributions seems to be quite restrictive. One could think that normal distributions should be more appropriate for this kind of model. In fact, with normal distributions, the different cases that made the analysis complex would disappear and it would be easier to formalize the process of information acquisition. However, the triple integration will probably not lead to closed form solutions. The qualitative results and especially the fact that inside information will become a disadvantage with a high number of bidders should be true for more general distributions.

The assumption of perfectly asymmetric information between banks is certainly unrealistic, but probably not more than the usual assumption of conditional independence of information. It is this assumption that allows to apply the simple dominance criterion as equilibrium concept. This means that, unlike for much of the common value auction theory, no far reaching assumptions on the players' rationality are required.

Note that there is a small change of the situation with respect to which the model is not robust: Imagine that the insider has a participation in the firms equity. In this case, he is interested in low interest rates if the competitor gives the credit and may therefore try to bid the interest rate down. Thus, the dominance criterion does not apply any more. An insider may want to stay in the auction even if his break-even rate is reached in order to further lower the interest rate. A precise solution in the open bid setting with affiliated values is, however, difficult to find. The idea will be further exploited in the next chapter using in the sealed bid approach from chapter 2.

Of course the application of this model is not restricted to the banking context. Asymmetrically informed bidders and affiliated values are features observed at many auctions. The "insider's curse" effect has not been identified in auction theory so far. Particularly interesting applications could be expected in the theory of share price behavior and market mircrostructure.

Chapter 4

Equity ownership of banks

4.1 Introduction

The participation of banks in the equity capital of a firm is one of the most controversial issues in banking regulation. The maximum amount of capital that a bank is allowed to hold in a non-financial firm differs widely between countries. Usually two types of restrictions can be found: limitations of the absolute level of equity investment in relation to a bank's capital and absolute limits on the participation in a single company. Whereas the first kind of restrictions are normally only relevant for very big firms, direct limitations of bank ownership are a severe limitation of the banks' scope of action. For example in Denmark and Norway banks may own up to 50% of a firm's voting stock, in Portugal 25%, in Finland and Ireland 10% and in Belgium, Japan and Sweden up to 5%. Since Italy has recently abandoned ownership restrictions, Australia is now the only OECD country which does not allow any investment in equity of non-financial firms. Similar to Germany, Austria, Greece, Spain and Turkey have no general restrictions on the percentage of a firm that a bank may own (OECD 1992).

The regulations differ considerably in detail. In several countries, exemptions form the restrictions can be granted by government agencies. Sometimes non-voting participations may exceed the listed percentage. In some countries, restrictions only apply to certain types of banks or to certain industries. For example in Turkey, banks are not allowed to take participations in real estate and building companies.

It is interesting to see that the possibility to own shares has nothing to do with the financial system being bank-dominated or market-based. For example, Italy has a very much bank-dominated financial system. Nevertheless, banks are not allowed to own shares, except for the purpose of underwriting. On the other hand. the UK has a market-centered financial system where in principle banks are allowed to own shares.

Two well-known extremes are Germany and the USA. Germany has a bank-centered financial system where banks are free to own non-financial firms. In the USA, the participation of national banks in a non-financial firm's capital had already been restricted in the 1863 National Banking Act. The 1933 Glass-Steagall Act (Roe 1990) then also eliminated the possibility for banks to deal in securities through affiliates. Finally, the Bank Holding Company Act of 1956 restricted bank holding companies to own less than 5% of a non-banking company.

Interestingly, both countries are thinking about changing their legislation. In Germany, the introduction of restrictions on bank ownership has been discussed in the Bundestag (Bundestag (1997)). In a recent study, the German Monopolkommission (1998) has recommended to limit bank participations to 5%. In the US, the Congress recently started its 10th attempt since 1933 to abolish the Glass-Steagall Act (Economist 1998).

What are the reasons for restricting bank ownership of firms? Several possible effects have been discussed: for example universal banking is thought to endanger the stability of the financial system, impair the investment bank activity and influence the lending activity of the bank.

This paper concentrates on the last aspect. The basic point is that the restriction of equity ownership by banks should rather be understood as part of the antitrust regulation than as prudential regulation. It is shown that an equity participation of one bank in a firm will lead to a reduced competition in the market for credits to that firm. This is in contradiction to the usual argumentation. So far, the impact of bank shareholding on the lending activities of the bank has been used as an argument in favor of universal banking. Most researchers seem to agree that bank holdings of equity tend to benefit the firms. Indeed, it is quite intuitive that the participation of a bank in the capital of the firm leads to an alignment of interests between the firm and the bank. A share-holding bank should be willing to finance riskier projects, obtain better information, and have more incentives for restructuring in case of distress. Neuberger and Neumann (1990) or John et alii. (1995) are presenting models in this spirit. Sometimes the alleged benefits of universal banking have led to a quite enthusiastic view of the German system (Calomiris (1994)).

The entirely positive view of the benefits of bank participations has never gained broad acceptance in Germany. An extensive and mainly critical discussion of the effects of bank influence can be found in Albach (1999). Whereas the benefits of a close relationship between banks and firms are generally recognized, suspicions of banks exploiting market power through share possessions have been expressed during the whole of the twentieth century[1]. Nevertheless only recently, after in 1995 a new law imposed the disclosure of participations of more than 5%, it has become possible to underline this suspicion with empirical evidence.

Several studies have estimated the influence of bank participation in a firms capital on the interest rate paid by German firms. Schmid (1996) uses a Herfind-

[1] Compare Haas (1995) for a historical overview.

ahl index to measure the concentration of banks in the firms' capital. He observes a significant interest rising effect of higher concentration [2]. Seger (1998) replicates these results using total bank participation as the explaining variable. In a nonlinear specification, he obtains decreasing interest rates for small participations and only for higher equity stakes a positive effect. Other studies provide complementary evidence. Albach (1998) and Schwiete/Wiegand (1997) observe that firms with higher bank participations have lower debt to equity ratios. The reason could be that debt is more expensive for firms with bank participation. Albach and Elston (1994) use a criterion that includes bank participation in order to define firms with close bank relationships. They cannot confirm that firms with close bank relationships are less liquidity constrained, what would be expected if asymmetric information was the driving force for financial constraints. Similar results have been obtained by Weinstein and Yafeh (1998) for Japan. They show that the cost of finance for banks with close bank ties is higher than that of their peers.

The existing economic models of bank participations cannot explain these observations as they do not allow for market power effects. Building on the models of the first chapter, banking competition is modeled as a common value auction with sealed bids. Thus it is possible to show that, indeed, the ownership of a fraction of the firms capital gives a bank market power in the competition for the firm's credits, which leads to higher cost of finance for the firm. In addition, more precise predictions about the relationship between bank participation and interest rates can be derived. If more than one bank has a participation in a firm, the interest rate depends on the asymmetry of the banks' participations. Identical stakes of several banks will not increase the interest rate but decrease it compared to the situation in which no bank owns a share of the firm. However, a situation in which both banks own equity stakes will not be an equilibrium of a game in which the banks buy shares before competing in the credit market.

The results rely on very basic payoff structures: a bank that owns shares of a company is not only remunerated by the interest payments, but gets a part of the firm's cash flows through dividends or increased market value of the shares. Hence it is more inclined to lower the interest rate than a non share owning competitor. However, if the bank is sure to give the credit, it still wants the firm's interest rate to be high because, unless it does not entirely own the firm, it recovers only a part of the cash flows through dividends. Only in case the competitor provides the loan, it will prefer low interest rates. The strategic situation is similar to that of a takeover auction when the bidders already own a fraction of the equity before the start of the bidding. These auctions have been recently studied by Burkart (1995) in a private value setting and by Bulow, Huang and Klemperer (1996) for common values. Whereas the price level rises with private values, it decreases

[2] It should be mentioned that Schmid does not attribute this effect to an increased market power of one bank.

on average with common values. The main source of uncertainty in banking is probably the credit risk which can be modeled as a common value. However, because in credit auctions, unlike in standard auctions, there is no lowest price (highest interest rate) for which the bidders are sure not to make losses, the model of Bulow, Huang and Klemperer is not directly applicable.

4.2 The Model

Two banks competing for giving a loan to a firm are considered. Both banks own fractions α_i, $i = 1, 2$ of the firm's equity. Without restriction of generality, it is assumed that $\alpha_2 \geq \alpha_1 \geq 0$.

Except for the equity participation of the bank, the model is exactly the same as in chapter 2. In the following the setup will be briefly recalled: The size of the loan is normalized to one. The firm has an investment project, which is going to succeed with probability λ. If the project succeeds, it yields a payoff $X > 1$. Hence the maximum interest rate that a project can support is $X - 1$. If the project is successful, the credit is paid back and the net profit is distributed to shareholders. With probability $1 - \lambda$, the project fails and has zero return. The assumption $\lambda X < 1$ is made, so that even if the banks ask the highest possible interest rate, it is not profitable to lend without additional information.

With probability q, both banks independently receive additional information about the quality of the firm. It is again assumed that, having received this information, they know with certainty whether the firm's project is going to succeed or not. As it is not profitable to lend without additional information or after having received negative information, banks only make an offer if they know the project is going to succeed.

This offer consists in the repayment b the firm has to make after the project has succeeded. The interest corresponding to this repayment is $b - 1$. It is supposed that both bids are made simultaneously and that the competitor neither observes the offered interest rate nor the fact that a bid has been submitted.

Again, the game has no equilibrium in pure strategies. If one bank were always bidding the same interest rate, the best response of the other bank would be either to slightly undercut this bid or to always bid the highest possible amount. Clearly in both cases the first bank's bid is not optimal.

Let $F_i : [1, X] \rightarrow [0, 1]$ be the equilibrium distribution functions for the bid b_i of bank i in the case it has received a positive signal. The expected profit of bank j, having received a positive signal, bidding b and having an equity participation of $\alpha_j \geq 0$ is:

$$\begin{aligned} \pi_j(b) &= (b-1)\left[q\left(1-F_i(b)\right)+(1-q)\right] \\ &\quad +\alpha_j\left[\int_1^b qF_i'(x)(X-x)\,dx+\left(q\left(1-F_i(b)\right)+(1-q)\right)(X-b)\right], \end{aligned} \tag{4.1}$$

The first term is the profit on the credit business $(b-1)$ multiplied by the probability of giving the loan. The second brackets contain the profit the bank will make on its equity participation. The integral term are the expected dividends in case the credit goes to the other bank and the other terms describe the dividends in case bank j wins the auction.

In a mixed strategy equilibrium, both banks have to choose the distribution of their bids such that the competing bank is indifferent between its possible bids. Otherwise it cannot be optimal for the bank to randomize over the bids. It is shown in the appendix C how this condition leads to a set of differential equations, which characterize the equilibrium.

Proposition 15 *The following distribution functions F_1, F_2 describe the unique equilibrium of the game:*

$$F_1(b) = \begin{cases} 0 & \text{for } b < 1 + (X-1)(1-q)^{\frac{1}{1-\alpha_1}} \\ \frac{1}{q} - \frac{1}{q}\left(\frac{(1-q)^{\frac{1}{1-\alpha_1}}(X-1)}{b-1}\right)^{1-\alpha_2} & \left[1 + (X-1)(1-q)^{\frac{1}{1-\alpha_1}}, X\right) \\ 1 & \text{for } b \geqq X, \end{cases}$$

$$F_2(b) = \begin{cases} 0 & \text{for } b < 1 + (X-1)(1-q)^{\frac{1}{1-\alpha_1}} \\ \frac{1}{q} - \frac{1}{q}\frac{(1-q)(X-1)^{1-\alpha_1}}{(b-1)^{1-\alpha_1}} & \left[1 + (X-1)(1-q)^{\frac{1}{1-\alpha_1}}, X\right) \\ 1 & \text{for } b \geqq X. \end{cases}$$

Proof. See appendix C ■

The higher is one bank's bidding density in a given area, the higher are the competitor's incentives to lower its bid. However, with equal bidding densities the incentives to lower the bid of the bank with the higher shareholdings are always bigger. Therefore, in order to keep the bank with the higher equity stake indifferent between its bids, the other bank has to bid with a lower density. Because the bank with the smaller stake cannot bid with lower density on the entire support, it has to put positive probability on the highest bid X.

Fig.4.1 shows the distribution functions of both bidders for $\alpha_1 = 0$, $\alpha_2 = 0.4$, $X = 1.2$, and $q = 0.7$. It is easy to see that the average bid of the bank with the higher shareholding is lower and that, in case both banks have received a positive signal, the bank with higher shareholding wins with a higher probability

How can this behavior be interpreted? In fact the bank with the smaller equity stake realizes that it is not worthwhile competing with the other bank if both have received a positive signal. Instead, it tries to extract the maximum profit in case the other bank has not obtained a signal by bidding the highest possible amount.

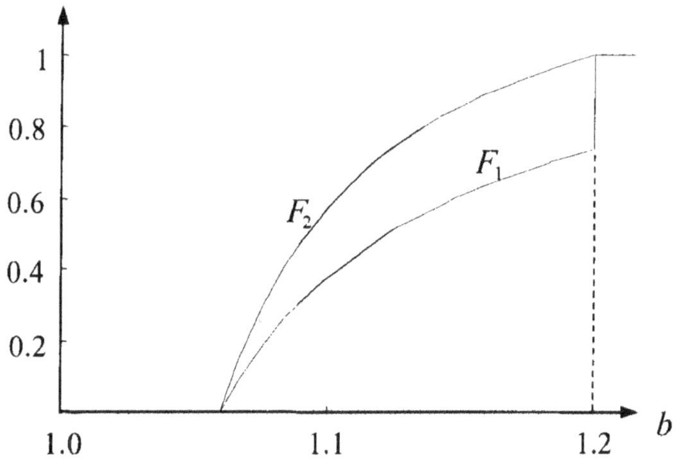

Figure 4.1: Bidding distributions for $\alpha_1 = 0$, $\alpha_2 = 0.4$, $X = 1.2$, $q = 0.7$.

The fact that the shareholding bank is able to bid more successfully than the non-shareholder does not yet allow to judge the overall effect of equity stakes on the average interest rate. Two countervailing forces determine the equilibrium interest rate: on the one hand banks with shareholdings are less averse to placing lower bids, as they are able to recover a part of their losses through the dividend. This will lead to a lower interest rate. On the other hand, however, banks with smaller or no shareholdings anticipate this and realize that it is not worthwhile competing with the shareholding bank. This may increase interest rates. In order to decide which one of these effects is dominant, the comparative statics of the equilibrium have to be derived.

First it is examined how an increase in the participation of bank 2, the bank with the bigger equity stake, influences the equilibrium outcome.

Proposition 16 *A bigger equity stake of bank 2 increases its profit from the credit business and does not decrease the profit of bank 1.*

Proof. Here the simple case $\alpha_1 = 0$ is demonstrated. The reader is referred to appendix C for the quite involved general calculations. With $\alpha_1 = 0$ the profit of bank 1 is easily obtained by inserting the bidding function $F_2(b)$ in 4.1 and multiplying with the probability of receiving a good signal:

$$\begin{aligned} \lambda q \pi_1(b) &= \lambda q (b-1)\left[q(1-F_2(b)) + (1-q)\right] \\ &= \lambda q (1-q)(X-1) \end{aligned} \tag{4.2}$$

This does not depend on α_2. Not surprisingly this is the same profit it would get by always bidding the highest possible amount and only winning the auction in case the competitor has not received any information. Otherwise, the bank would have no reason to bid the highest bid with a positive probability.

In order to calculate the profit of bank 2's credit department, $F_1(b)$ has to be plugged into 4.1 leaving out the dividend term:

$$\lambda q (b-1)\left[q(1-F_1(b)) + (1-q)\right] = \lambda q (b-1)^{\alpha_2}\left[(1-q)(X-1)\right]^{1-\alpha_2} \tag{4.3}$$

As bank 2's credit department is not indifferent between all possible bids, it is necessary to integrate over the distribution of b in order to obtain the expected profit from the credit business:

$$\Pi_2 = \int_{1+(X-1)(1-q)}^{X} \lambda q (b-1)^{\alpha_2}\left[(1-q)(X-1)\right]^{1-\alpha_2} F_2'(b)\, db \tag{4.4}$$

$$= \int_{1+(X-1)(1-q)}^{X} \lambda q (b-1)^{\alpha_2}\left[(1-q)(X-1)\right]^{1-\alpha_2}\left[\frac{(1-q)(X-1)}{q(b-1)^2}\right] \tag{4.5}$$

$$= \lambda \int_{1+(X-1)(1-q)}^{X} \left[\frac{(1-q)(X-1)}{b-1}\right]^{2-\alpha_2} db \tag{4.6}$$

$$= \frac{\lambda}{1-\alpha_2}\left[(1-q) - (1-q)^{2-\alpha_2}\right](X-1) \tag{4.7}$$

In order to know how this profit changes with the size of the equity stake, the derivative of this term with respect to α_2 is taken:

$$\frac{\partial}{\partial \alpha_2}\Pi_2 = \frac{(1-q)^{2-\alpha_2}}{(1-\alpha_2)^2}\left[(1-q)^{-(1-\alpha_2)} - 1 + \ln(1-q)^{(1-\alpha_2)}\right]\lambda q (X-1) > 0 \tag{4.8}$$

As $\ln(x) > 1 - \frac{1}{x}$,this is always positive. ■

Perhaps surprisingly, with nonzero α_1, the profit of bank 1 also strictly increases with a higher participation of bank 2. In our setting, the generated surplus is fixed since the probability of a credit does not change. Hence, the increased profit of the banks can only come through an increase in the average interest rate at the disadvantage of the firm's shareholders. More precisely, if Π_1^C and Π_2^C are the expected profits of bank 1 and 2's credit departments, the average interest rate a good firm has to pay conditional on getting a credit is $\frac{\Pi_1^C+\Pi_2^C}{[1-(1-q)^2]\lambda}$. As the sum of the banks' profits increases it is clear that:

Corollary 6 *The average interest rate the firm has to pay increases with the size of bank 2's equity stake.*

This already provides a strong argument in favor of restricting bank participation in a firm's equity. However, it could be argued that the starting situation of the game is not an equilibrium of a more complete game. It could be suspected that if both banks can decide whether to acquire a participation in the firm, the outcome will be symmetric and no bank will keep an advantage. The next proposition shows that this is not the case.

Proposition 17 *A bigger stake of bank 1 in the firm's equity decreases the profit of both banks credit departments.*

Proof. See appendix C

This means that a game in which both banks were allowed to buy shares can only have asymmetric equilibria with one bank owning no shares. A situation where both banks own shares cannot be an equilibrium, since selling the (weakly) smaller participation to the bank with the bigger equity stake would increase both banks' profits from the credit business.

It is not easy to determine whether the effect of proposition 16 or the effect of proposition 17 is stronger, i.e. to derive what happens if both stakes are increased simultaneously. In fact, only in the symmetric situation an unambiguous effect can be obtained:

Proposition 18 *If both banks hold symmetric equity stakes, the average interest rate decreases with increasing participation.*

This shows that it does not make much sense to look for a relationship between total bank participation and average interest rates. If for reasons exogenous to this model several non-cooperatively acting banks own similar shares, the interest rate will not increase but decrease. Perhaps this explains why Seger (1998) observed for low participations decreasing interest rates. I has been shown that equity stakes of several banks are not an equilibrium outcome. Hence symmetric participations have probably not been chosen strategically and tend therefore to be small.

4.3 Conclusion

Allowing banks to own equity stakes in firms will reduce the competition in the credit market and drive up the cost of debt for the firms. This explains recent empirical results obtained with German and Japanese data. Whereas the original motivation for the Glass Steagall Act and much of the regulation in other countries seems to have been unjustified (Benston (1995), Rajan/Kroszner (1994),

Puri (1994)), this effect provides a new argument for limiting bank ownership in non-financial firms.

The interest-raising effect relies on very basic payoff structures. The mechanism works with non-voting equity as well as with ordinary shares. The result is only driven by the fact that a bank having an equity participation is interested in high interest rates as long as it is giving the credit itself, but prefers low interest rates in case the other bank is giving the loan. The interest-raising effect has been derived in a possibly unfavorable setting: there is no danger of a winner's curse which drives the similar result in the model of Huang, Bulow and Klemperer (1996). Because the ex ante profitability of a firm from the sample is assumed to be negative, nobody makes a bid without knowing that the project is good. Introducing a winner's curse by extending the analysis to the case of lending with inconclusive signals would probably further increase the interest-raising effect. In addition, in this case the shareholding banks will be probably more lenient lenders than the non-shareholding.

There are several other mechanisms, not captured by the model, which could contribute to the interest raising effect in reality. For example, the bank could simply influence the management's choice of a credit through its votes on the firm's board. Another reason for reduced credit market competition could be that the share owning banks have access to better information. It has been shown in chapter 2 how this can increase the cost of finance for the firm. Shareholding could be used as a signal to outside banks that information acquisition about this firm is not worthwhile.

Chapter 5

Usury and Credit Rationing

5.1 Introduction

The simple model of monitored finance which is constructed in the first part of this chapter helps to explain several features of a bank-firm relationship. However, the prinicipal intention of this chapter is to give a new explanation for credit rationing by banks. The attitude of banks with respect to credit risk has puzzled economists for a long time. The interest rates banks quote do not differ very much among firms of different riskiness. For example Machauer and Weber (1998) analyze the dependence of interest rates on a bank's own internal rating of borrowers. In their sample, the average interest rate difference between borrowers in the best and the worst of five risk classes is only 1.2%. Given that the best class is defined as "good or very good creditors" whereas the worst class consists of borrowers which are "very much in danger of default" these interest differences are much smaller than the interest rate spreads on corporate bonds of comparable ratings. In addition, the results of Petersen and Rajan (1994) or Blackwell and Winter (1997) seem to indicate that the variations of the interest rate can rather be explained with differences in bargaining power or credit market competition than with the riskiness of a loan.

Instead of increasing the interest rate in order to compensate for the risk, banks seem to turn down credit for perfectly profitable but risky firms, or restrict the amount of credit to a firm with a profitable project. This behavior is referred to as credit rationing. Note that not every borrower is being rationed if he doesn't receive financing, even if he is willing to accept a higher interest rate. The bank's refusal to lend may be justified, because the borrower's project has negative value. Following Keeton (1979) two sorts of rationing can be distinguished: the rationing of the number of loans but not of the loan size is called type I rationing, whereas an inefficient reduction of the loan size is referred to as type II rationing. It is evident that rationing can have considerable consequences for the development of an economic system, especially in countries where banks are the dominating source of finance. Innovative projects with high returns but also high risk will not be financed, the efficiency of the allocation of capital will decrease.

Credit rationing behavior was already postulated in the 1950s by Roosa (1951) and Scott (1957) who argued that monetary policy operates through the availability of loans rather than the interest rate. However, it was only at the end of the 1970s that attempts were made to explain credit rationing as an equilibrium phenomenon. Jaffee and Russel (1976), Baltensperger (1978) and Keeton (1979) have clarified the concept and pioneered the asymmetric information approach which culminated in the influential paper by Stiglitz and Weiss (1981). The key effect in the Stiglitz/Weiss model is that interest can only be paid in case a project succeeds. As firms do not pay for the downside risk, the expected interest rate for a risky project is lower than the one for a safe project. Hence, if the bank increases the interest rate, entrepreneurs with safe projects will not apply for finance any more whereas entrepreneurs with risky projects of the same expected payoff will still be willing to take the credit. Alternatively, safe entrepreneurs may try to increase the riskiness of their projects. Because of these adverse selection and moral hazard effects, higher interest rates may eventually lead to a lower profit for the bank. The bank will be forced to quote a low interest rate for which the demand for capital is higher than the supply. Therefore, some borrowers will be rationed.

A different line of research, including Townsend (1979), Gale/Hellwig (1985) and Williamson (1986) and (1987) obtains credit rationing as a side effect of theories initially developed for explaining the optimality of debt contracts under asymmetric information. Rationing is here the consequence of ex post monitoring or bankruptcy costs. In the Williamson (1986) model, banks are reluctant to ask high, risk-adjusted interest rates because this increases the probability of failure and therefore the bankruptcy costs.

The Stiglitz/Weiss approach has been criticized from several sides. Riley (1987) has observed that rationing of this type should not be frequent if banks can distinguish the riskiness of firms. DeMeza and Webb (1987) show that the results of Stiglitz/Weiss rely on a special assumption on the distribution of risk in the population of firms. Given the theoretical weaknesses, it is not surprising that empirical studies have not succeeded in observing the effects implied by the asymmetric information approach to credit rationing. Berger and Udell (1992) show that loan rate stickiness does not change in a way consistent with asymmetric information theories of credit rationing. Cressy and Toivanen (1997) develop a test in order to directly decide between theories using symmetric versus asymmetric information. Their results suggest that information is imperfect, but symmetric.

In this chapter, it is argued that the asymmetric information approach, while valuable for commercial bonds and other types of arm's length debt, is misleading when used to analyze relationship banking which is typical for continental Europe and especially Germany. As argued by Albach (1997), on the contrary, the good information of the bank about the firm and the mutual trust is the key to understand the advantages of bank financing and the problems that may arise.

Monitored finance The idea that finance provided by a well informed bank may increase the efficiency of firms has been used by several authors in recent papers. Carminal and Matutes (1997) assume that after having invested in monitoring, banks can directly influence the decisions of the firms and will do this in a socially beneficial way. In Holmström and Tirole (1997) monitoring by a bank alleviates moral hazard by decreasing the private benefits of risky behavior. In this chapter, a different mechanism is proposed which explains why banks can provide more efficient corporate governance than bond markets .

It is argued that the decisive difference between a bank loan and a bond is that, whereas the interest rate and the duration of a bond are fixed once for all (except the violation of prespecified covenants), most credit contracts contain an unspecific "Material Adverse Clause" (MAC) which gives the bank the discretion to call back the credit or increase the interest rate whenever this seems appropriate. In fact, this seems to be more common for relationship based banking than for economies where arm's length finance prevails. It is anecdotal evidence that credit contracts in Germany are far less detailed and give the bank much more discretion than in the United States. Typically in the United States the "Material Adverse Clause" only applies to loan commitments, whereas in Germany it is included in almost all credit contracts.

The economic function of the MAC cannot be analyzed if the firm is modeled as a one shot random variable like in most of the asymmetric information literature. The MAC makes only sense if distress situations do not occur suddenly but can be anticipated by a bank. This idea is formalized in the model presented in section 5.2. The models closest to this setting are Rajan (1992) and Rajan (1995). Rajan focuses on the incentives to obtain efficient liquidation and gives the bank the right to change the financing conditions only at the end of a loan contract or if prespecified covenants are violated. In contrast to this, here it is assumed that the bank is free to choose the timing and the type of its action, which is definitely more typical for European-style relationship banking. It then does not make much sense to assume that the bank will lose money when liquidating. In general, a closely monitoring bank will step in before this state of distress is reached.

The discretion to raise the interest rate gives the bank the possibility to provide efficient incentives to the entrepreneur. The intuition of this result is very simple: the entrepreneur is interested in increasing the riskiness of his project as long as the interest rate he has to pay is fixed. However, if the increased risk is observed by the bank and it adapts the interest rate, the entrepreneurs incentives to increase risk disappear.

Whereas the good and possibly exclusive information of the bank increases the efficiency, it may have negative effects on the competitiveness of the market. There are two reasons why banks that have not monitored the firm will incur a higher credit risk. They are not able to anticipate risky situations and react appropriately and this in turn leads to inefficient incentives for the firms which further increases credit risk. Hence banks that have not monitored the firm will not be able to provide credit even if they can observe the interest rate offered by the monitoring bank. Monitoring on the other hand is not profitable for an outsider as he will never recoup his monitoring costs. Hence the flexibility of the credit contract, together with informational advantage of the bank, explains the possibility of ex post rent extraction.

Reputation Of course, giving the bank the possibility to change the credit contract in a situation where outside banks cannot step in, creates a problem of moral hazard at the side of the bank. Banks may simply claim that the firm's rating has deteriorated. Asking a higher interest rate is only one of the possibilities to extract additional rents. Another practice which becomes increasingly popular in Germany with banks diversifying into consulting is to send the banks consulting or restructuring team to the firm and charge a hefty bill for these "services".

The most efficient solution for the moral hazard problem of the bank would be to directly condition the contract on the credit risk. However, credit risk is probably a variable that is not verifiable in court. It can only be tried to capture credit risk as closely as possible with verifiable covenants. However, as there are always unforeseen contingencies, covenants are never perfectly correlated with the credit risk. Efficiency can only be approached by more and more complex contracts.

This chapter argues that relationship finance prevailing in continental Europe uses another mechanism. Typically changes in the terms of the credit contract are left very much at the discretion of the lender. It seems that the banks' reputational concerns prevent them from misusing their power. In fact, discretionary contracts and reputation are complementary features. Boot at alii (1993) have shown that banks relying on a reputational mechanism may even leave contractual features discretionary that could in principle be enforced by courts. Using discretionary contracts instead of legally enforceable ones, facilitates the development of reputation.

The importance of reputation in bank-firm relationships has been recognized by several authors. Whereas Diamond (1989) and (1991) analyzes the impact of the debtor's reputation on his cost of finance and project choice, Sharpe (1990), Cemmanur and Fulghieri (1994) and Dinc (1997) have pointed out the importance of the bank's reputation. Closest to the model used in this chapter is Sharpe (1990) who also assumes that the banks acquire a reputation not to exploit informational rents. Cemmanur and Fulghieri (1994) focus on the decision of the financier to liquidate a firm in financial distress or renegotiate the credit. They show that reputational concerns will incite banks to make more costly evaluations of the firm than bondholders in order to take the correct decision. Dinc (1997) considers the incentives of banks to observe discretionary loan commitments.

Similarly to these papers, it is shown in section 5.3 how in a repeated game banks will treat customers correctly in order not to lose profits from future financing relationships. However, the main focus is not the reputation mechanism itself but its possible disadvantages. It is demonstrated how the necessity to maintain a good reputation may lead to type I credit rationing.

The intuition is very simple: some of the actions which are appropriate for a firm in difficulties can as well serve to exploit perfectly healthy firms. If firms are not able to distinguish between a bank exploiting healthy firms and a bank recovering its credits in a bankrupt firm, they may misinterpret a justified action as usurious practices. In order to avoid being suspected to be a usurer, banks may prefer not to finance too many risky firms.

In a recent paper Carey et alii (1996) provide strong empirical support for this explanation. They compare the differences in the credit portfolios of finance companies and banks and observe that finance companies lend to significantly riskier borrowers. Carey et alii (1996) suggest that this is because finance companies care less about their reputation. However, they do not explicitly model this, nor recognize the possible implications of reputation on credit rationing. Their paper shows that some risky firms which have been refused by reputation-based banks have the possibility to get arm's length finance from a finance company. However, a lot of firms which could in principle be financed with monitored bank debt may not be able to get arm's length finance because then the Stiglitz/Weiss moral hazard problem becomes relevant. In this sense, the basic mechanism of credit rationing used in this chapter is the same as in Stiglitz/Weiss. However, the chapter does not apply it to banks but only to arm's length finance. The proposed explanation of credit rationing neither needs the assumption of asymmetric information between a bank and a firm, nor risk aversion of the bank. Note that whereas the bank is not risk averse in the economic sense of having a concave utility function, it behaves risk adverse in a common language sense, as it refuses to finance risky but profitable firms.

A key insight of the model is that it may not be sufficient to look at a single bank-firm relationship if one wants to explain why a firm is rationed credit. In section 5.3, it will be shown that rationing depends on the risk profile of the entire economy. Identical banks may be willing to finance the identical firm in one particular economic environment but not in another.

Note that the danger of usurious practices explains the astonishing popularity and the success of non profit-maximizing financial institutions especially in developing, but also in industrialized countries. Allan and Gale (1995) point out that a substantial fraction of total banking assets in Germany are held by savings banks. They are not profit maximizing entities, but were originally set up to provide credit to the poor and finance local and regional investments and continue at least partially to pursue these objectives. Only 26% of total banking assets in Germany are held by private commercial banks whose attachment to shareholder value maximization is at least doubtful, given their managerial entrenchment documented by the Monopolkommission (1998).

The model also shows that it is in principle not unreasonable to subsidize risky but promising firms with credits from a public bank such as the Kreditanstalt für Wiederaufbau (KfW), the European Investment Bank (EIB) or the European Bank for Reconstruction and Development (EBRD). A non profit maximizing public bank doesn't need to sustain a reputation. It can provide first best incentives, because the firm has no reason to fear that it will be exploited. Of course it might be difficult to regulate the public bank appropriately. Without having to maximize profits it may have the tendency to be too lenient.

5.2 Monitored finance

5.2.1 The model

A lending relationship extends over three points in time. At $t = 0$, firms need an investment which is normalized to 1, in order to generate a success cash flow of X at $t = 2$. However, at $t = 1$, with probability $1 - p(e)$, depending on the invested effort of the entrepreneur $e \in [0, \infty]$, the firm enters in a risky situation.

This is described by assuming that the probability of success decreases considerably to q. The function $p : [0, \infty] \to (0, 1)$ is assumed to be increasing and concave i.e. $p'(e) > 0$, $p''(e) < 0$. Having invested the monitoring costs c, a bank is able to observe this change in the credit quality of the firm. If the credit contract contains a "materially adverse" clause, it can react by either liquidating the firm and recovering $L < qX$, or increasing the interest rate. Figure 5.1 summarizes the structure of the model.

First best solution: The overall surplus created by the project is

$$[q(1 - p(e)) + p(e)] X - 1 - e, \tag{5.1}$$

hence the first best effort e^* is characterized by:

$$\frac{\partial}{\partial e} p(e_{FB}) = \frac{1}{(1 - q) X} \tag{5.2}$$

In the sequel, p^* will be used as shorthand notation for $p(e^*)$. The first best surplus plus the effort costs will be denoted by $\Pi_{FB} := [q - qp_{FB} + p_{FB}] X - 1$.

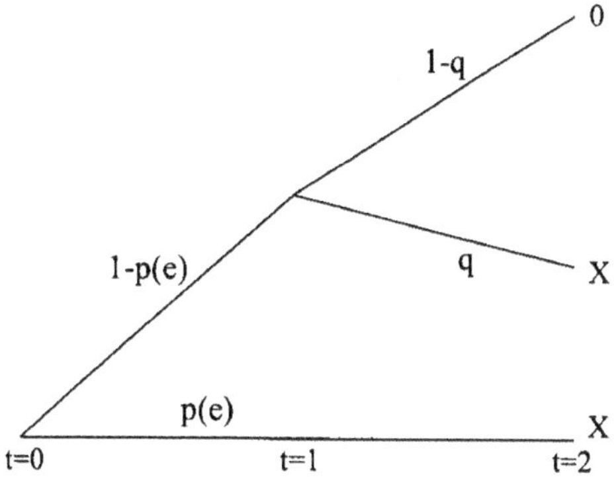

Figure 5.1: Time structure of a lending relationship

Limited liability and contingent contracting: Assume that the deterioration of the borrowers credit rating is a contractible contingency or that the bank is a welfare maximizing institution. Then the first best solution can be implemented by conditioning the contract on the credit risk. Denoting the repayment in the riskless case by b_s and by b_r, the repayment in the risky situation, the borrower will chose his effort such that:

$$\max_{e\in[0,\infty]} \left[(1-p(e))\,q\,(X-b_r)+p(e)\,(X-b_s)-e\right] \tag{5.3}$$

$$\Rightarrow \frac{\partial}{\partial e}p(e)=\frac{1}{(X-b_s)-q\,(X-b_r)}=\frac{1}{X(1-q)+qb_r-b_s} \tag{5.4}$$

Of course the bank has to make non-negative profits. Its participation constraint is:

$$p(e)\,b_s+(1-p(e))\,qb_r\geq 1 \tag{5.5}$$

Obviously, the firm owner has higher incentives to invest effort, if the interest rate increases in the risky state. He will choose the first best effort if:

$$\frac{1}{X(1-q)+qb_r-b_s}=\frac{1}{(1-q)\,X} \tag{5.6}$$

$$\Rightarrow qb_r=b_s, \tag{5.7}$$

i.e. if the increase in the interest rate exactly offsets the risk.

Arm's length contracts: If the bank is not monitoring or the credit contract does not contain a "materially adverse" clause, the repayments must be equal in both states, i.e. $b_s = b_r = b$. In this case the firm will invest effort such that

$$\Rightarrow \frac{\partial}{\partial e} p\left(e_{AL}\right) = \frac{1}{(X - b)(1 - q)} > \frac{1}{(1 - q) X} \tag{5.8}$$

If the bank makes zero profit, this implies an interest rate of,

$$b = \frac{1}{\left[p_{AL} + \left(1 - p_{AL}\right) q\right]} > 1 \tag{5.9}$$

Riskier projects require a higher interest rate, which in turn decreases the entrepreneur's incentives to invest effort. Therefore, arm's length finance may not be feasible from a certain riskiness on. The surplus created with arm's length contracts plus the effort costs will be denoted as $\Pi_{AL} := \left[q - qp_{AL} + p_{AL}\right] X - 1$.

Usurious practices: A firm that does not trust a bank has lower incentives to invest effort in its project. Assume that the firm assigns a probability of ρ to the possibility of being exploited by the bank. In this case, the bank increases the interest rate and deprives the firm of its entire surplus.

The effort chosen by the firm depends on the expectation of being exploited:

$$\max_{e \in [0,\infty]} \left[(1 - p(e)) q (X - b_r) + p(e)(X - b_s)\right](1 - \rho) - e \tag{5.10}$$

$$\Rightarrow \frac{\partial}{\partial e} p(e) = \frac{1}{\rho (1 - \rho)\left[(X - b_s) - q (X - b_r)\right]} = \frac{1}{\rho (1 - \rho)\left[X (1 - q) + q b_r - b_s\right]} \tag{5.11}$$

If X is big enough or q small enough, first best investment can still be achieved by raising b_r resp. lowering b_s until

$$(1 - \rho)\left[X (1 - q) + q b_r - b_s\right] = (1 - q) X \tag{5.12}$$

$$\Leftrightarrow \frac{\rho}{1 - \rho} X (1 - q) = q b_r - b_s \tag{5.13}$$

However, not for all ρ, the bank's participation constraint may hold. If first best investment is implemented then the bank's profit π can be calculated as:

$$p_i^* b_s + (1 - p_i^*) q b_r - 1 = \pi. \tag{5.14}$$

Equations 5.13 and 5.14 can be solved for $b_r$and b_s.

$$\begin{aligned} b_s &= \pi + 1 - \frac{\rho}{1-\rho} X (1-q)(1-p_i^*) \\ b_r &= \frac{1}{q}\left(\pi + 1 + \frac{\rho}{1-\rho} X (1-q) p_i^*\right) \end{aligned} \tag{5.15}$$

With decreasing ρ the bank's profit has to decreases, otherwise the firms will turn to arm's length financing:

$$(1-\rho)(\Pi_{FB} - \pi) - e^* = \Pi_{AL} - e_{AL} \tag{5.16}$$

$$\Rightarrow \pi = \Pi_{FB} - \frac{\Pi_{AL} + e_{FB} - e_{AL}}{1-\rho} \tag{5.17}$$

The critical exploitation probability is:

$$\hat{\rho} = 1 - \frac{\Pi_{AL} + e_{FB} - e_{AL}}{\Pi_{FB}}. \tag{5.18}$$

For $\rho > \hat{\rho}$, trustworthy banks will not be able to offer a profitable credit contract.

Liquidation: It is assumed that liquidation is inefficient and that a bank liquidating the firm has not increased the interest rate beforehand. The liquidation value of the firm is L. If a bank is always liquidating in the risky state, the entrepreneur's maximization problem has the following form:

$$\max_{e \in [0,\infty]} \left[(1-p(e)) \max[(L-b), 0] + p(e)(X-b) - e\right] \tag{5.19}$$

under the participation constraint:

$$p(e)\, b + (1-p(e)) \min[L, b] \geq 1 \tag{5.20}$$

$$\Rightarrow \frac{\partial}{\partial e} p(e) = \frac{1}{(X-b) - \max[(L-b), 0]} \tag{5.21}$$

Liquidation will induce first best effort if:

$$\frac{1}{(1-q)X} = \frac{1}{(X-b) - \max[(L-b), 0]} \tag{5.22}$$

$$\Rightarrow qX = \max[L, b] \tag{5.23}$$

Inefficiency implies $qX > L$, as long as $b < qX$ liquidation will induce excess effort. Note that for low L the project may not have a positive value, so it may not be worthwhile to finance a project if they have to be liquidated in risky situations.

Competition The existence of monitoring costs explains in a straightforward way why competition to take over a credit at $t = 1$, will be weak. Without monitoring, outside banks cannot make a profitable offer. If they offer an interest rate between $b_s - 1$ and $b_r - 1$, they will only get the risky credits and make losses. If they offer an interest rate $1 + r > b_r$ they will get no credits, or in the limit case not make profits. Of course, monitoring is not worthwhile either, as they will never be able to recoup the monitoring costs, which the other bank has already sunk.

If initially several trustworthy banks are competing for the credit, the interest rate a firm has to pay in the safe state would be equal to the refinancing cost of the bank. This means that, in this case, credit market competition boils down to a pure private value auction. Of course, this is because we have excluded any common value factors, such as the possibility of immediate distress. However, the model shows that private value components may play a more important role than has been realized in most auction models of banking competition.

Note, however, that the efficiency-increasing role of flexible financial contracting is robust with respect to and may even be increased with intermediary competition. If we assume that $c = 0$, the intermediary interest rates the firm would be able to obtain in a competitive credit market without informational asymmetry would be the one that incites the first best effort. Indeed, as two monitoring and not colluding banks are sufficient for perfect competition, this may explain why most firms choose to maintain two bank relationships. Von Thadden (1994) has worked out this idea in a slightly different model, assuming that banks will deliberately encourage competition and introduce "second sourcing" in order to credibly commit to low interest rates. In a less competitive environment it seems more probable that the firms are the ones which do everything to prevent a hold up by a bank. They may establish "second sourcing" for credits by allocating credits to a second bank despite less favorable conditions. Empirically it has been shown by Harhoff (1998) that the majority of German firms receives credit from two banks. In the sequel, we do not pursue these ideas further, but simply assume a monopolistic bank.

5.2.2 Discussion

Despite its simple structure, several features of bank finance can be explained with this model. It shows how the flexible clauses of a credit contract together with a well informed bank, which is acting in the interest of the borrower, can increase the efficiency of finance. As a side result, it explains why initially interest rates are not risk adjusted but rather reflect the refinancing conditions of the bank. This also explains why banks have insufficient incentives to screen the borrower's quality very much before making a credit offer. Indeed, all they have to know about the borrower is that his project has a positive expected value. The precise risk of the firm is not very relevant. Anyway the initial interest rate does not depend on the firm's risk, as long as the credit risk is not too high.

Whereas there is no need for a bank to sink monitoring costs before the firm has accepted the credit offer, monitoring is necessary once the credit contract is signed. In contrast to the Sharpe (1990) and Rajan (1992) models, it is not simply assumed that the banks acquire information during the lending relationship, but the information acquisition is modeled explicitly. It is shown that the banks will rather collect information that enables them to anticipate and deal with distress situations than information about the quality of the borrower. Therefore, the credits of one bank cannot been simply be transferred to another bank at the same interest rate. This is an explanation for Fishers (1933) observation, that the bankruptcy of banks may have severe consequences on the availability of credits for firms. Furthermore it provides a straightforward and robust explanation for the possibility of ex post rents.

Of course the model is very stylized. It could be made more realistic by reintroducing the possibility of a nonanticipated failure of the firm. This would add a common value component to the bank's evaluation of the credit. The banks would have to compensate this kind of risk with a risk premium form the beginning of the relationship on. However, as this premium would only include the imminent credit risk it will probably be quite small.

A richer version of this model could be constructed with techniques derived from option pricing. In reality new information influencing the credit risk of a bank arrives continuously in time, so that the credit risk follows a jump diffusion process. The bank can hedge against the continuous components of this process but is forced to compensate the jumps by risk premium. Like in Holmström/Milgrom (1987) the contract giving the entrepreneur optimal incentive to influence this process could be analyzed. It could be shown that highly volatile businesses make frequent monitoring necessary. The high monitoring costs could be a reason not to finance these projects or to prefer non-monitored incentives through equity finance.

5.3 Credit Rationing and Reputation

This part will analyze what happens if the bank is not a non-profit maximizing organization and if the credit risk is not contractible. It will be demonstrated how the phenomenon of credit rationing can be explained by the necessity of the banks to maintain a good reputation. In fact, there are two distinct mechanisms which will lead to credit rationing under different circumstances. One will rather be observed in stable economies, the other will be prevailing in risky environments.

The intuition of the direct mechanism is easy to understand. If the contract allows the bank to increase the interest rate, firms are afraid to be exploited by the bank. They may observe the past behavior of the bank before applying for a credit. If they are not able to distinguish between justified and non-justified renegotiations of the interest rate, they may prefer not to go to a bank which has renegotiated too many credits. Anticipating this, honest banks may be reluctant to finance profitable but highly risky firms, for which the credits have to be renegotiated quite often. Banks have to trade off the losses from not financing risky firms today against the losses from its diminished reputation in the future. This mechanism becomes especially relevant if there are many firms with low risk and only a few risky firms. Banks will be reluctant to endanger their reputation in such a situation. They will not gain very much from financing the few high-risk firms, but may loose a lot if the low-risk firms do not come back in the future. This could be the reason why credit rationing is a concern in mature and stable economies like Germany. There the number of firms which can be financed almost without incurring risk is substantially larger than the amount of profitable but highly risky firms.

However, there is a second mechanism which indirectly leads to credit rationing in unstable environments. If the sample of firms is rather risky, credits have to be renegotiated very often. This means that, even at honest banks, a lot of renegotiations can be observed. In this case, firms will not take a renegotiation very seriously. This in turn will deteriorate the incentives of opportunistic banks to behave correctly, the reputation mechanism will break down. Only by rationing credit and selecting a small sample of firms with low risk, honest banks are still able to maintain a reputation. If honest banks are not able to decrease the credit risk in their sample below a critical value, no reputational equilibrium may be sustainable and monitored finance may become impossible altogether. Contrary to the first scenario, this is a situation most probably encountered in highly unstable economies, for example in developing countries or economies in transition.

5.3.1 The model

There are two main methods of modelling reputation. One possibility is to construct an infinitely repeated game, where trigger strategies can be used to enforce a behavior which would be no equilibrium of the stage game. This approach has been used in most of the banking literature (e.g. Sharpe (1990), Bagnioli (1992), Dinc (1997)). However, for this kind of models it is important that the firms perfectly observe the banks action, as only one deviation or wrong observation leads to a breakdown of the reputational equilibrium. Since imperfect observability is crucial for explaining rationing, the approach is not very well suited for the purpose of this chapter.

Therefore, in this chapter, reputation is modeled in a finitely repeated game in the tradition of Kreps et alii. (1982), where the presence of irrationally honest types of players induces opportunistic players to mimic honest behavior. Benabou and Laroque (1992) or Fudenberg and Levine (1992) have shown that reputation is sustainable in these games even with imperfect observability of one player's actions. Since two repetitions of the stage game are already sufficient to observe credit rationing, the model is limited to this case. As usual in reputation games, more repetitions would decrease the fraction of honest and dishonest players necessary for sustaining the equilibrium, but not change the qualitative effects.

Assumptions Each period, a continuum with mass one of the firms described in section 5.2 arrives. The firms differ in the ex ante probability with which they enter in the risky state. They are ordered by riskiness and indexed by a parameter $x \in [0,1]$. A higher index x implies a higher risk. The ex ante probability with which a firm x enters in the risky state, even the owner invests the first best effort e^*, is denoted by $p(e^*, x)$. The fraction $\lambda(x)$ of firms that enter in a risky state, if all firms with $x \in [0, \hat{x}]$ are financed, can then be calculated as:

$$\lambda(\hat{x}) = \frac{1}{\hat{x}} \int_0^{\hat{x}} p(e^*, x)\, dx. \tag{5.24}$$

It is assumed that, whereas monitored finance is possible, the moral hazard problem is too serious for arm's length finance to be feasible. Hence the firms have no access to the bond market. The firms have to get financed by a monopolistic bank which may be of one of three possible types. With probability $\boldsymbol{\alpha}$, the bank is **trustworthy** and only renegotiates the interest rate if firms have really entered in a risky situation. With probability $\boldsymbol{\beta}$, the bank is an **opportunist** who renegotiates the interest rate whenever this can be done without loosing future customers and with probability $\boldsymbol{\gamma} = 1 - \alpha - \beta$, the bank is a **usurer** who always exploits a big fraction $\frac{1}{2} < \lambda_u < 1$ of its customers. All types of banks strategically choose the fraction $\hat{x}$ of firms they want to finance. Only the opportunistic banks also decide strategically about the amount y of firms they want to exploit.

It is assumed that an opportunistic bank exploits successful and risky firms with equal probability. Exploitation will yield more for firms which are going to succeed than for firms which are going to enter in a risky situation. However, on expectation an exploiting bank will extract the first best surplus Π_{FB}. If opportunistic banks have decided not to exploit a firm they do not skim of the entire surplus, even when the firm enters in a risky situation. They only adapt the interest rate to the new risk.

There are two generations of firms. The firms of the second generation observe the amount of $\hat{x}$ firms financed, but have imperfect information about the behavior of the bank in the first period. This is modeled by assuming that they randomly draw two firms out of the sample financed by the bank in the first period. They are able to see if the interest rates of these two firms have been renegotiated, but they do not know whether this was justified or not. Depending on their observation, they decide to go to the bank or not to finance the project. If the risk of being exploited is too high, eventually the expected return will be smaller than the effort they have to invest and they will give up their project.

Setting $\Pi_{Al} = e_{AL} = 0$ in equation 5.17 gives the profit which a non-exploiting bank is able to make, depending on the exploitation probability ρ expected by the firms.

$$\pi = \Pi_{FB} - \frac{e_{FB}}{(1-\rho)} \tag{5.25}$$

The critical exploitation level for which honest banks will not be able to offer a credit contract any more is $\hat{\rho} = 1 - \frac{e_{FB}}{\Pi_{FB}}$. For ρ higher than $\hat{\rho}$, the probability of not being exploited is so small that investing the necessary effort is not worthwhile for the firm.

The banks' strategies: In principle the bank decides each period about x and y if it is an opportunistic type or only about x if it is the honest type or the usurer. However, some of the choices can be trivially determined. As there is no reason to maintain a reputation after the last period of the game, all types of banks will finance the entire sample of the firms in the last period. In addition, opportunistic banks will exploit all firms.

As the second period's firms are able to perfectly observe the amount x of firms financed in the first period, the usurers and the opportunistic types have to exactly mimic the decision which is optimal for the honest bank. We can therefore reduce the decision variables of the banks to the amount x of firms financed in by the honest bank in the first period and the fraction y of firms exploited by the opportunistic types in the first period.

The firms' strategies: The first period's firms only have the choice to go to the bank or not. The information set of the second period's firms has the three possible elements "no renegotiation ", "one renegotiation", "two renegotiations". Hence they can choose among the three strategies: "go to every bank", "go to a bank which has less than two renegotiations", "only go to a bank if you do not observe a renegotiation".

Summarizing this discussion, after having eliminated the trivial choices, the game has the following structure.

Structure of the game:

1. In the first period, a continuum of firms seek finance.
1.1. Each firms decides whether to apply for credit or not.
1.2. The bank decides about the fraction x of firms it is willing to finance.
1.3. An opportunistic bank decides about the fraction y of firms to exploit.
1.4. The stage game takes place for every firm.

2. In the second period, a new continuum of firms arrives
2.1. Every firm observes credit renegotiations for two old firms.
2.2. Every firm decides whether to apply for credit or not.
2.3. The stage game takes place.

It might be useful to recall the stage game:

1. The firms invest in effort.
2. The state of nature materializes for each firm.
3. Banks renegotiate interest rates.
4. The firms' outcome materializes and credit is paid back.

Two things can destroy a reputation mechanism in this setting: low profit margins and a too risky sample. If the gains from exploiting firms are very big compared to the rents earned with trustworthy behavior, the banks will have no incentives to sustain a reputational equilibrium. This is a well known effect which has been derived for example by Dinc (1997).

The second reason why the reputation mechanism can not be sustainable is the presence of too much good, but risky firms. If a lot of firms enter into the risky state, even trustworthy banks will have to renegotiate the interest rate quite often. This may make it difficult for a firm to distinguish between a usurer and a trustworthy bank. As a consequence, the firms may not react very severely upon observing a renegotiation of the interest rate. They will be willing to go to a bank even if they know that it has previously renegotiated credits, because they know that with a high probability is was justified. But then, the opportunistic banks will realize that they can exploit at least some firms without losing a large number of customers in the future. Finally, if firms anticipate that opportunistic players will exploit them, they will prefer not to apply for credit. In such a situation, credit rationing can help to establish a reputational equilibrium, if it sufficiently reduces the average risk in the sample. We summarize this discussion in a proposition:

Proposition 19 *In situations where no reputational equilibrium is sustainable when all firms are financed, credit rationing can make lending possible if it reduces the risk of the bank's credit portfolio sufficiently.*

Proof. See Appendix D ∎

If the risk cannot be lowered sufficiently, finance may become impossible altogether. In this case only non profit-maximizing banks will still be able to provide credit.

In the above case, the trustworthy banks had to ration credit because otherwise no financing would have been possible. However, it may even be optimal to ration credit in situations where an unrationed reputational equilibrium would be sustainable. If a reputational equilibrium is sustainable, in the first period trustworthy banks earn $\pi_1 = \pi(\lambda_u \gamma)$ and in the second period $\pi_2 = \pi(\gamma + \beta)$ for each firm they finance.

If they finance a fraction x in the first period, in the second period with probability $1 - (1 - \lambda_t(x))^2$ a firm observes a renegotiation of the interest rate and prefers not go to the bank. Therefore, the two period profit of the bank will be:

$$x\pi_1 + (1 - \lambda_t(x))^2 \pi_2 \tag{5.26}$$

Depending on the form of $\lambda_t(x)$, the maximum of this function may lie in the interior of $(0, 1]$. In this case, the firm will ration credit in the first period. The first order condition

$$\pi_1 - 2(1 - \lambda_t(x)) \lambda_t'(x) \pi_2 = 0 \tag{5.27}$$

is necessary, but not sufficient. The interior maximum may not be the global one. Financing all firms in the first period may still be more profitable. The following proposition summarizes this result.

Proposition 20 *Credit rationing occurs if the function* $x\pi_1 + (1 - \lambda_t(x))^2 \pi_2$ *has an interior maximum on* $(0, 1]$.

It is not easy to give more intuitive criteria to determine whether credit rationing will occur. A very simple example may help to reduce the number of parameters and facilitate the understanding. Assume $\pi_1 = \pi_2$ and that the sample only consists of a fraction ξ of riskless firms and $1-\xi$ firms with the probability $\hat{p}$ of entering in the risky state.

Corollary 7 *Credit will be rationed if* $\hat{p} > \frac{1}{1+\sqrt{\xi}}$.

Proof. See Appendix D ■

Credit rationing is more probable, the higher the fraction of riskless firms and the higher the risk of the risky firms. Intuitively, the first order condition $\pi_1 - 2(1 - \lambda_t(x)) \lambda_t'(x) \pi_2 = 0$ holds if $\lambda_t'(x)$ is big and $\lambda_t(x)$ small, i.e. if by financing more firm one adds considerable risk to a relatively safe sample. At the same time, x should not be too small because then it might be more profitable to finance all risky firms instead of a few safe ones, i.e. then the maximum may be on the border.

5.3.2 Discussion

In this chapter, only some of the consequences of the reputation-based approach have been explored. Some additional conclusions and possible extensions will be discussed in this paragraph.

For example, the model could be easily extended to explain excess liquidation by banks. Banks which are concerned about their reputation may prefer to liquidate risky firms instead of raising the interest rates, even if liquidation is inefficient. Contrary to raising the interest rate, liquidating is not profitable when applied to a healthy firm. It is only rational if the firm is really in a difficult situation. Therefore, liquidation cannot be misinterpreted as an attempt to exploit informational rents. Liquidation will facilitate the establishment of reputation, but not solve the rationing problem. Now, the risky firms may be rationed because they are unprofitable if they are liquidated in a difficult situation.

Another simple extension would be the introduction of arm's length finance. Similar to the Rajan (1992) paper, the model could then be used to analyze the choice of a firm between bank and market finance.

Instead of assuming that there are different types of banks, the decision of the bank to exploit customers could also be related to the financial situation of the bank. Banks which are in a liquidity crisis may have to exploit customers in order to meet their obligations, or may just be unwilling or unable to provide credit when necessary. For example, prudential regulation limits the extend to which credits can be given by a bank in financial difficulties. Banks which finance a lot of risky projects may be more likely to enter in a liquidity crisis if they are not perfectly diversified. Peek and Rosengreen (1997) or Houston, James and Marcus (1996) provide evidence that the constraints of prudential regulations can be binding and that consequently credit supply changes with a bank's financial situation. Therefore, firms may prefer more stable banks. Of course, they also prefer stable banks because it may be difficult and expensive to obtain monitored credit from another bank if the own housebank has gone bankrupt.

5.4 Conclusion

Using the simplest possible mathematical specifications this chapter has tried to formalize how a close financing relationship, which gives a bank a lot of bargaining power may increase efficiency but creates at the same time a commitment problem for the bank, which can only be solved by a possibly inefficient reputation mechanism.

The model has implications for bank management as well as for policy making. The most important lesson is definitely that bank finance is fundamentally different from market-based financing through bonds and equity. Management and valuation techniques derived from the capital markets should not be simply applied to bank lending. Short-term profit oriented strategies, which neglect reputational concerns may have severe consequences when applied to a bank. It is of course always possible to cash in the reputation and briefly increase the profit of a bank, but this will have irreversible consequences for its future profits.

In particular, reputation can not only be destroyed by non-cooperative behavior. Frequent restructurations of the lending techniques and frequent changes in the bank personnel may also make the firms feel that the past may not be a good indicator for the bank's future behavior. Of course, changing a bank's image may help to increase a bank's reputation if it is already bad.

The model also underlines the importance of the loan officers and the informal information they posses. They should not only be considered as a costly distribution system of the bank's product. In particular, they, rather than the bank as a whole, may be the ones in which a firm trusts. Even if their information collection tasks could be carried out more efficiently by computerized scoring and credit supervision techniques, loan managers may be necessary for establishing reputation. This also confirms Mitusch's (1995) result that the monetary incentives of loan officers should not incite them to extract short term profits from the firm.

It has already been mentioned that the model can be used to justify the existence of non-profit maximizing banks, specialized in lending to small, innovative and risky firms. Another policy recommendation that can be justified with the model is to implement better transparency and disclosure regulations. This will facilitate the establishment of reputation and increase the competition of banks in the intermediary stage of a lending relationship.

An interesting side effect of the model is that the reputation of banks in general affects the feasibility of trustworthy behavior. If the probability of usurious behavior becomes too high, trustworthy banks will not be able to offer profitable contracts. This could also help to explain why the economic recovery after banking crises may be difficult. Tirole (1996) shows that once the reputation of a group has been destroyed, for example by a random shock, it may be impossible to rebuild reputation. Therefore, if in a crisis a large number of banks had to exploit their customers in order to avoid bankruptcy, it will be difficult to re-establish the reputation-based relationship-banking.

Bibliography

[1] Albach, H. and J. Elston (1995), Bank Affiliation and Firm Capital Investment in Germany, *Ifo-Studien, 41*:3-16.

[2] Albach, H. (1997), Eine lebendige Theorie der Unternehmung, *Frankfurter Allgemeine Zeitung,* December, 11th 1997, 21-22.

[3] Albach, H. (1998), Bankbeziehungen und Wettbewerbsfähigkeit, *Zeitschrift für Betriebswirtschaft,* Ergänzungsheft :1-25.

[4] Albach, H. (1999), Corporate Governance, *Jaakko Honko Lectures*, Helsinki School of Economics and Busines Administration, HeSE print, Helsinki.

[5] Allen, F. A. and D. Gale (1995), A Welfare Comparison of Intermediaries and Financial Markets in Germany and the US, *European Economic Review,* 39:179-209.

[6] Amable, B., J. B. Chatelain and O. DeBandt (1997), Stability versus Efficiency of the Banking Sector and Economic Growth, *Banque de France*, Mimeo.

[7] Ang, J. and T. Richardson (1994), The Underwriting Experience of Commercial Bank Affiliates prior to the Glass Steagall Act, *Journal of Banking and Finance*, 18:351-395.

[8] Aoki, M. and S. Dinc (1997), Relational Financing as an Institution and its Viability under Competition, Stanford University, Mimeo.

[9] Arrow, K. (1964), The Role of Securities in the Optimal Allocation of Risk Bearing, *Review of Economic Studies*, 32(2):91-96.

[10] Baglioni, A. (1992), Bank-Firm Customer Relationships as Dynamic Game Equilibria, *Rivista Internazionale di Scienze Economice e Commerciali,* 39, (10/11): 943-958.

[11] Benston, G. (1995), The Origins of and the Justification for the Glass-Steagall Act, in: A.Saunders and I. Walter eds.: Universal Banking: Financial System Design Reconsidered, Chicago: Irwin.

[12] Berger, A. and G. Udell (1992), Some Evidence on the Empirical Significance of Credit Rationing, *Journal of Political Economy,* 100, (5):1048-1076.

[13] Benabou, R. and Guy Laroque (1992), Using Privileged Information to Manipulate Markets: Insiders Gurus and Credibility, *Quarterly Journal of Economics.*

[14] Bhattacharya, S. and A. Thakor (1993), Contemporary Banking Theory, *Journal of Financial Intermediation,* 3:2-50.

[15] Boot, A. W. and S. I. Greenbaum (1993), Reputation and Discretion in Financial Contracting, *American Economic Review, 83, (5):1165-1183.*

[16] Broecker, T. (1990), Credit-Worthiness Tests and Interbank Competition, *Econometrica, 58*:429-452.

[17] Bulow, M., J. Huang and P. Klemperer (1996), Toeholds and Takeovers, CEPR Discussion Paper, 1996.

[18] Bundestag (1997), Protokoll der 74. Sitzung des Rechtsausschusses, Rechtsausschuß, Deutscher Bundestag, Bonn.

[19] Burkart, M. (1995), Initial Shareholdings and Overbidding in a Takeover Contest, *Journal of Finance, 50*:1491-1515.

[20] Calomiris, C. (1994), The Cost of Rejecting Universal Banking: American Finance in the German Mirror, in N. Lamoreaux and D. Raff, eds.: Coordination and information: Historical perspectives on the organization mirror, 1870-1914. Chicago: University of Chicago Press.

[21] Caminal, R. and C. Matutes (1997), Bank Solvency, Market Structure and Monitoring Incentives, Barcelona, Institut d'Analisi Economica, Mimeo.

[22] Caminal, R. and C. Matutes (1997) Can Competition in the Credit market be Excessive? CEPR Discussion Paper No. 1725.

[23] Campbell, C. M. and D. Levin (1997), Can the Seller Benefit from an Insider in Common Value Auctions? Ohio State University, Mimeo.

[24] Carey, M., M. Post and S. A. Sharpe (1996), Does Corporate Lending by Banks and Finance Companies Differ? Evidence on Specialization in Private Debt Contracting, Federal Reserve Board, Mimeo.

[25] Carlin, W. and C. Mayer (1998), Finance, Investment and Growth, Oxford, mimeo.

[26] Chan, Y.-S., S. Greenbaum and A. V. Thakor (1986), Information Reusability, Competition and Bank Asset Quality, *Journal of Banking and Finance,* 10:243-253.

[27] Chemmanur, T. J. and P. Fulghieri (1994), Investment Bank Reputation, Information Production and Financial Intermediation, *Journal of Finance, 49:57-79.*

[28] Chiappor, P. A., Perez-Castillo and F. Verdier (1995), Spatial Competition in the Banking System, Localization, Cross-Subsidies and the Regulation of Interest Rates, *European Economic Review,* 39, (5):889-919.

[29] Cordella, T. and E. Levy Yeyati (1998), Financial Opening, Deposit Insurance and Risk in a Model of Banking Competition, CEPR Discussion Paper No. 1939.

[30] Cressy, R. and O. Toivaven (1997), Is there Adverse Selection in the Credit Market? Helsinki School of Economics, Mimeo.

[31] Debreu, G. (1959), The Theory of Value, New York, Wiley.

[32] Diamond, D. (1984), Financial Intermediation and Delegated Monitoring, *Review of Economic Studies,* 40:393-414.

[33] Diamond, D. (1989), Reputation Acquisition in Debt Markets, *Journal of Political Economy,* 97:828-862.

[34] Diamond, D. (1991), Monitoring and Reputation: The choice between bank loans and directly placed debt, *Journal of Political Economy,* 99:688-721.

[35] Dinc, S. (1997), Bank Reputation, Bank Commitment and the Effects of Competition in Credit Markets, Mimeo, University of Tokio.

[36] Economist (1998), American Financial Regulation: The Great Escape, *The Economist,* March14th -20th: 98.

[37] Elberfeld, W. and E. Wolfstetter (1999), A dynamic Model of Bertrand Competition with Entry, *International Journal of Industrial Organization 17*(4):531-526.

[38] Fischer, K. (1990), Hausbankbeziehungen als Instrument der Bindung zwischen Banken und Unternehmen. Eine theoretische und empirische Analyse, Univ. Diss, Universität Bonn. I. Fisher (1933), The debt-deflation theory of great depressions, *Econometrica,* 1:337-357.

[39] Freixas, X. and J.-C. Rochet (1997), Microeconomics of Banking, MIT Press, Cambridge, Massachusetts.

[40] Schnitzer, M. (1999), Enterprise Restructuring and Bank Competition in Transition Economies, *Economics of Transition* *7*(1):133-155.

[41] Fudenberg, D. and D. K. Levine (1992), Maintaining a Reputation when Strategies are Imperfectly Observed, *Review of Economic Studies,* 59:561-579.

[42] Gale, G. and M. Hellwig (1985), Incentive Compatible Debt Contracts: The One-Period Problem, *Review of Economic Studies.*

[43] Garella, P.G. (1996), Informational Rents in Interbank Competition, *Recherches Economiques de Louvain,* 62, (1):3-20.

[44] Gehrig, T. (1996), Market Structure, Monitoring and Capital Adequacy Regulation, *Swiss Journal of Economics and Statistics,* 132, (4/2):685-702.

[45] Gehrig, T. (1998), Screening, Cross-Border Banking and the Allocation of Credit, Universität Freiburg, Mimeo.

[46] Goldsmith, R. (1969), Financial Structure and Developement, Yale University Press, New Haven.

[47] Gorton, G. and A. Winton (1998), Banking in Transition Economies: Does Efficiency Require Instability? *Journal of Money; Credit and Banking,* 30:621-655.

[48] Haas, J. (1994), Der Anteilsbesitz der Kreditwirtschaft an Nichtbanken, Univ. Diss., Universität des Saarlandes.

[49] Harsanyi, J. (1973), Games with Randomly Distributed Payoffs: A new Rationale for Mixed Strategy Equilibrium Points, *International Journal of Game Theory,* *2*:1-23.

[50] Hausch, D. and L. Li (1993), A Common Value Auction Model with Endogenous Entry and Information Acquisition, *Economic Theory,* 3:315-334.

[51] Hirshleifer, D. and D. Kent (1996) A theory of costly sequential bidding, University of Chicago, Mimeo.

[52] Holmström, B. and P. Milgrom (1987), Aggregation and Linearity in the Provision of Intertemporal Incentives, *Econometrica,* 55:303-328.

[53] Holmström, B. and J. Tirole (1997) Financial Intermediation, Loanable Funds and the Real Estate Sector, *The Quarterly Journal of Economics,* 112, (3):663-691.

[54] Houston, J. and C. James (1996), Banking Relationships, Financial Constraints and Investment: Are Bank Dependent Borrowers More Financially Constrained? University of Florida, Mimeo.

[55] Houston, J., C. James and C. Marcus (1996), Adverse Selection Problems and Bank Lending: Evidence from Internal Capital Markets, University of Florida, Mimeo.

[56] Innes, R.D.(1990), Limited liability and incentive contracting with ex-ante action choices, *Journal of economic theory.* 52, (1):45 - 67.

[57] Jayaratne, J. and P. E. Strahan (1996), The Finance Growth Nexus: Evidence from Bank Branch Deregulation, *The Quarterly Journal of Economics*, 111:639-670.

[58] Jensen, M. and N. Meckling (1976), The Theory of the Firm: Managerial Behavior, Agency Costs and Ownership Structure, *Journal of Financial Economics, 3, (4):305-360.*

[59] John, K., T. A. John and A. Saunders (1994), Universal Banking and Firm Risk-Taking, *Journal of banking & finance*,18:307 - 323.

[60] Kagel, J. H. and D. Levin (1998), Common Value Auctions with Insider Information, forthcoming in *Econometrica.*

[61] Kamecke, U. (1997) Dominance or Maximin: How to solve an english auction, Humboldt-Universität zu Berlin, Mimeo.

[62] Kanniainen, V. und R. Stenbacka (1998), Project Monitoring and Banking Competition under Adverse Selection, Wissenschaftszentrum Berlin für Sozialforschung, Discussion Paper 97,23.

[63] King, R. and R. Levine (1993), Finance and growth: Schumpeter might be right, *The quarterly journal of economics 108*:717 - 737.

[64] Klein, M. (1971), A Theory of the Banking Firm, *Journal of Money Credit and Banking, 3*:205-18.

[65] Kroszner, R. and R. Rajan (1994), Is the Glass Steagall Act justified? A Study of the US Experience with Universal Banking Before 1933, *American Economic Review, 84, (4)*:810-823.

[66] Levine, R. and S. Zervos (1998), Stock Markets, Banks and Economic Growth, *American Economic Review,* 88, (3):537-557.

[67] Matthews, S. (1984), Information Acquisition in Discriminatory Auctions, in *Bayesian Models in Economic Theory,* edited by M. Boyer and R.Kilstrom, Elsevier Science Publisher B.V., Chapter 11, 182-207.

[68] Milgrom, P. and R. J. Weber (1982), A theory of auctions and Competitive Bidding, *Econometrica,* 50:1089-1123.

[69] Mitusch, K. (1998), "Information Acquisition by Middlemen" Discussion Paper, Freie Universität Berlin.

[70] Mondigliani, F. and M. Miller (1958), The Cost of Capital, Corporation Finance and the Theory of Investment, *American Economic Review* 68:261-297.

[71] Monti, M. (1972), Deposit, Credit and Interest Rate Determination under Alternative Bank Objectives, In: Mathematical Methods in Investment and Finance, edited by G.P. Szego and K. Shell, Amsterdam North-Holland.

[72] Monopolkommission (1998), Ordnungspolitische Leitlinien für ein funktionsfähiges Finanzsystem, Sondergutachten 26, Nomos Verlagsgesellschaft, Baden-Baden.

[73] Neuberger, D. and M. Neumann (1991), Banking and Antitrust: Limiting Industrial Ownership by Banks, *Journal of Institutional and Theoretical Economics*, 188-206.

[74] OECD (1992), Banks under Stress, *Organisation for Economic Co-operation and Development*, Paris.

[75] OECD (1989), Competition in Banking, *Organisation for Economic Co-operation and Development*, Paris.

[76] Pagano, M. (1993), Financial Markets and Growth: An Overview, *European Economic Review,* 37:613-622.

[77] Peek, J. and E. S. Rosengren (1997), Collateral Damage: Effects of the Japanese Commercial Real Estate Collapse on Credit Availability and Real Activity in the Unted States, Federal Reserve Bank of Boston, Mimeo.

[78] Petersen, M. and R. Rajan (1995), The Effect of Credit Market Competition on Lending Relationships, *The Quarterly Journal of Economics, 110,* (2):407-443.

[79] Petersen, M. and R. Rajan (1994), The Benefits of Lending Relationships: Evidence from Small Business Data, *Journal of Finance* 49:3-37.

[80] Porter, M. E. (1992), Changing the Way America Invests in Industry, Council of Competitiveness Washington D.C., Mimeo.

[81] Puri, M. (1994), The Long Term Default Performance of Bank Underwritten Securities Issues. *The Journal of Banking and Finance, 18*:397-418.

[82] Rajan, G. (1992), Insiders and Outsiders: The Choice Between Informed and Arms-length debt, *Journal of Finance,* 45:1367-1400.

[83] Rajan, G. (1995a), Convenants and Collateral as Incentives to Monitor, *Journal of Finance,* 50:1115-1147.

[84] Rajan, R. (1995b), The Entry of Commercial Banks into the Securities Business: A Selective Survey of Theories and Evidence, University of Chicago, Mimeo.

[85] Rajan, R. and L. Zingales (1998), Financial Dependence and Growth, *American Economic Review,* 88, (3):559-586.

[86] Riordan, M.H. (1993), Competition and Bank Performance: A theoretical Perspective, In: C.Mayer and X. Vives, editors, *Capital Markets and Financial Intermediation*, Cambridge University Press: 328-343.

[87] Roe, M. J. (1990), Political and Legal Restraints on Ownership and Control of Public Companies. *Journal of Financial Economics, 27:*7-41.

[88] Rousseau, P. L and P. Wachtel, (1998), Financial Intermediation and Economic Performance: Historical Evidence from Five Industrialized Countries, *Journal of Money Credit and Banking,* 30, (4):657-678.

[89] Ruckes, M. (1998), Bank Competition, Credit Standards and Corporate Conservatism, Working Paper Universität Mannheim.

[90] Salop, S. (1979), Monopolistic Competition with Outside Goods, *Bell Journal of Economics,* 10, (1):141-156.

[91] Schmit, F. A. (1996), Beteiligungen deutscher Geschäftsbanken und Corporate Performance, *Zeitschrift für Wirtschaft und Sozialwissenschaften, 116*:273-310.

[92] Schumpeter, J. (1911), Theorie der Wirtschaftlichen Entwicklung, Reprint of the 4th edition, Ducker&Humblot , Berlin,1993.

[93] Schwiete, M. and J. Weigand (1997), Bankbeteiligungen und das Verschuldungsverhalten deutscher Unternehmen, *Kredit und Kapital, 30*:1-34.

[94] Seger, F. (1997), Banken, Erfolg und Finanzierung: Eine Analyse für deutsche Industrieunternehmen. Univ. Diss., Mannheim.

[95] Sharpe, S. (1990), Asymmetric Information, Bank Lending and Implicit Contracts: A Stylized Model of Customer Relationships, *Journal of Finance, 45:1069-1087.*

[96] Kannianinen, V. and R. Stenbacka (1998), Project Monitoring in Lending Markets with Adverse Selection, Swedish School of Economics, Helsinki, Mimeo.

[97] von Thadden, E. L. (1994), The Commitment of Finance, Duplicated Monitoring and the Investment Horizon, Universität Basel, Mimeo.

[98] Tirole, J. (1996), A Theory of Collective Reputations (with Appliations to the Persistence of Corruption and to Firm Quality), *Review of Economic Studies*, 63:1-22.

[99] Smith, T. (1998), Banking Competition and Macroeconomic Performance, *Journal of Money; Credit and Banking*, 30:793-815.

[100] Townsend, R. (1979), Optimal Contracts and Competitive markets with Costly State Verification, *Journal of Economic Theory, 21:265-293.*

[101] Tröge, M. (1997), Banking Competition as Mixed Private and Common Value Auction, WZB Discussion Paper FS IV 97-30.

[102] Tröge, M. (1998), Closed Bid Auctions with Common and Private Value Components, Wissenschafszentrum Berlin, Mimeo.

[103] Weinstein, D. E. and Y. Yafeh (1998), On the Costs of a Bank-Centered Financial System: Evidence from the Changing Main Bank Relations in Japan, *Journal of Finance,* 53, (2):635-672.

[104] Williamson, S. D. (1986), Costly Monitoring, Financial Intermediation and Equilibrium Credit Rationing, *Journal of Monetary Economics, 18:159-79.*

[105] Williamson, S. D. (1987), Costly Monitoring Equilibrium Loan Contracts and Equilibrium Credit Rationing. *Quarterly Journal of Economics,* 102:135-45.

[106] Yanelle, M. O. (1997), Banking Competition and Market Efficiency, *Review of Economic Studies,* 64:215-239.

[107] Yanelle, M. O. (1989), The Strategic Analysis of Intermediation, *European Economic Review,* 33:57-84.

Appendix

Appendix A

Proofs for Chapter 2

Proof of Proposition 1, 2, 1 and 2

The proofs of the propositions 1, 2, 1 and 2 to are all very similar. In order to verify that the distribution functions 2.9, 2.10,2.11 and 2.12 constitute an equilibrium it has to be shown that bank with a given information is indifferent on the support and makes lower profits from bidding outside the support. The general procedure will be sketched for bank 2 in case c) of proposition 1, the other cases can be treated accordingly.

i) inconclusive signal: If bank 2 has received an inconclusive signal. it is supposed to bid with the distribution 2.12 on the support $\mathcal{I}_4$. It has to be checked that it is indifferent between any bids on this interval. The profit of bank 2, having received an inconclusive signal and bidding on $\mathcal{I}_4$ can be calculated by plugging the functional forms of H_1,and. F_1 into equation 2.2:

$$\begin{aligned}
\pi_i^0(b) &= -(1-\lambda)q_1 + (\lambda b - 1)(1-q_1)\left[1 - \mu_1\frac{1}{\mu_1}\left[1 - \frac{(1-\lambda)q_2}{(b\lambda - 1)(1-q_1)}\right]\right] \\
&= -(1-\lambda)q_1 + (\lambda b - 1)(1-q_1)\left[\frac{(1-\lambda)q_2}{(b\lambda - 1)(1-q_1)}\right] \qquad \text{(A.1)} \\
&= (1-\lambda)(q_2 - q_1).
\end{aligned}$$

As required, this does not depend any more on b. In addition this is the postulated equilibrium profit, which proves proposition 2.

The above calculation holds for all b. This means that if bank 1 would bid on $\mathcal{I}_3$ with the same functional form as on $\mathcal{I}_4$, bank 2 would also be indifferent on $\mathcal{I}_3$. However for $b \in \mathcal{I}_3$ the bid distribution function H_1 is bigger than the functional form of H_1 on $\mathcal{I}_4$. Therefore the probability that bank 1 is winning is higher on $\mathcal{I}_3$, hence bank 2's profit is lower than if it was bidding on the upper interval $\mathcal{I}_4$. It is easy to see that bidding on $\mathcal{I}_1$ and $\mathcal{I}_2$ is still less profitable for bank 2.

ii) good signal: The profit of Bank 2, having received a good signal and bidding on $\mathcal{I}_3$ is

$$\pi_i^g(b) = (b-1)\left[(1-q_1)\left[1-\frac{b}{(b-1)}+\frac{1-\lambda q_1}{(b-1)(1-q_1)\lambda}\right]\right] \quad (A.2)$$

$$= \frac{1-\lambda}{\lambda}. \quad (A.3)$$

If it is choosing bids on $\mathcal{I}_2$, it is not bidding against bank 1 if bank 1 has not received a signal (because $H_1(b) = 0$), but against bank 1 if it has received a good signal. The profit is

$$\pi_i^g(b) = (b-1)\left[q_1\left(1-\frac{\lambda b-1}{\lambda q_1(b-1)}\right)+(1-q_1)\right] \quad (A.4)$$

$$= (b-1)q_1 - \frac{\lambda b-1}{\lambda}+(b-1)(1-q_1) \quad (A.5)$$

$$= \frac{1-\lambda}{\lambda}. \quad (A.6)$$

Hence bank two is indifferent on $\mathcal{I}_2 \cup \mathcal{I}_3$. For bids in $\mathcal{I}_4$, the same argument as before can be applied. The prolongation of the functional form of H_1 on $\mathcal{I}_3$ into $\mathcal{I}_4$ is smaller on $\mathcal{I}_4$ than the actual definition of H_1 on $\mathcal{I}_4$. Bank two would be indifferent for the functional form from $\mathcal{I}_3$, hence it looses with the actual definition of H_1 on $\mathcal{I}_4$. ■

Proof of Proposition 3

Case a) bidding only with good signals: In this case the payoff functions 2.24 and 2.25 take the form:

$$\Pi_1(q_1,q_2) = \lambda q_1(X-1)(1-q_1)-\gamma q_1^2, \quad (A.7)$$

$$\Pi_2(q_1,q_2) = \lambda q_2(X-1)(1-q_1)-\gamma q_2^2. \quad (A.8)$$

Solving this game leads to

$$q_1 = \frac{(X-1)\lambda}{\gamma+2(X-1)\lambda},$$

$$q_2 = \frac{(X-1)\lambda}{\gamma}\left[1+\frac{(X-1)\lambda}{\gamma+2(X-1)\lambda}\right].$$

We have $q_2 - q_1 = \left[\frac{(X-1)\lambda}{\gamma}+1\right]\left[1+\frac{(X-1)\lambda}{\gamma+2(X-1)\lambda}\right] > 0$, thus the solution is consistent with the condition $q_2 > q_1$.

The condition $q_1 = \frac{(X-1)\lambda}{\gamma+2(X-1)\lambda} > \frac{\lambda X-1}{\lambda(X-1)}$ has to hold
This is true for small enough γ

$$0 < \gamma < \lambda\frac{2(X-1)-\lambda(X^2-1)}{\lambda X-1}.$$

Case b) only bank 1 is bidding with good and bad signals In this case the equations 2.24 and 2.25 do not really constitute a game:

$$\Pi_i(q_i) = q_i(1-\lambda) - \gamma q_i^2$$

The firms' profit only depends on their own decision. But then the solution has to be symmetric which is in contradiction to being in situation b).

Case c) bidding with good and bad signals Now the banks' profit 2.24 and 2.25 can be written as:

$$\begin{aligned}
\Pi_2(q_2) &= \lambda q_2\left(\frac{1}{\lambda}-1\right) + (1-q_2)(q_2-q_1)(1-\lambda) - \gamma q_2^2, \\
\Pi_1(q_1) &= \lambda q_1\left(\frac{1}{\lambda}-1\right) - \gamma q_1^2.
\end{aligned}$$

For $\gamma > \frac{1}{2}(1-\lambda)$,this game has the equilibrium actions:

$$\begin{aligned}
q_1^* &= \frac{1-\lambda}{2\gamma}, \\
q_2^* &= \min\left\{\frac{(1-\lambda+4\gamma)(1-\lambda)}{(1-\lambda+\gamma)4\gamma}, 1\right\}.
\end{aligned}$$

We have $q_2^* - q_1^* = \frac{(1-\lambda)(\lambda+2\gamma-1)}{4\gamma(1-\lambda+\gamma)} > 0$. The equilibrium profits are

$$\begin{aligned}
\Pi_2(q_2^*) &= \frac{\left((1-\lambda)^2+8\gamma^2\right)(1-\lambda)^2}{16\gamma^2(1-\lambda+\gamma)}, \\
\Pi_1(q_2^*) &= \frac{(1-\lambda)^2}{4\gamma}.
\end{aligned}$$

■

Proof of Proposition 5 and 6

The proof is analogous to the proof of 1 in section A

Case a) bidding only with good signals Inserting the bidding distribution 2.42 of the $n-1$ competitors in 2.37, and observing that $H(b) = 0$, we obtain as the profit of a bank, having received a good signal and bidding in the interval $\left[\frac{1}{\lambda}, X\right]$:

$$\begin{aligned}
\pi^g(b) &= (b-1)\left[1-qF(b)-(1-q)\hat{\mu}_n H(b)\right]^{n-1} \\
&= (b-1)\left[(1-q)\left(\frac{X-1}{b-1}\right)^{\frac{1}{n-1}}\right]^{n-1} &\text{(A.9)}\\
&= (X-1)(1-q)^{n-1}. &\text{(A.10)}
\end{aligned}$$

Hence his ex ante expected profit is:

$$\pi = \lambda q (X-1)(1-q)^{n-1}$$

Bidding lower than $\frac{1}{\lambda}$,will decrease the profit as it will not increase the probability of winning compared to bidding exactly at $\frac{1}{\lambda}$, but decrease the interest rate in the case of winning.

Case b) bidding for good and inconclusive signals: the indifference of a bank having received a good signal can be calculated as above. Plugging the equilibrium distributions 2.43 and 2.44 of the competitors into the profit function 2.38, shows that a bank having received an inconclusive signal is indifferent on the support and making there zero profits. It will make losses from bidding below $\hat{b}_n$.

■

Proof of Proposition 10 and 9

Define

$$\frac{v-1}{r} = x. \tag{A.11}$$

The profit of a bank j having the refinancing rate v and bidding b is:

$$\begin{aligned}
\pi_j(b,v) &= (b-v)\left[q(1-F_i(b))+(1-q)\right] &\text{(A.12)}\\
&= (b-v)\left[q\left(1-F_v\left(b_i^{-1}(b)\right)\right)+(1-q)\right].
\end{aligned}$$

where $F_i(b)$ is the distribution of the competitor's bids, b_i his bidding function and

$$F_v(v) = \begin{cases} 0 & \text{for } v \leq 1, \\ \frac{1}{r}(v-1) & \text{for } 1 < v \leq 1+r, \\ 1 & \text{for } v > 1+r. \end{cases} \tag{A.13}$$

the distribution function of the private values. The random variable v can be obtained from a random variable x, which is uniformly distributed on $[0,1]$ with the transformation $v = 1 + rx$. The bidding strategies will be calculated as functions $b : [0, 1] \to [1, X]$ of x instead of v which will simplify taking the limit for $r \to \infty$.

Maximizing the bank's profit we obtain

$$\frac{\partial}{\partial b}\pi_j(b, v) = \left[1 - qF_x\left(b_i^{-1}(b)\right)\right] - (b - v)\left[qF_x'\left(b_i^{-1}(b)\right) b_i^{-1\prime}(b)\right] = 0. \quad (A.14)$$

The refinancing costs v are replaced with $1 + rx$, and x with $b_j^{-1}(b)$. Assuming the equilibrium is symmetric, the differential equation for the inverse bidding function $\sigma = b_i^{-1} = b_j^{-1}$, $\sigma : [1, X] \to [0, 1]$, can be written down

$$[1 - qF_x(\sigma(b))] - (b - (1 + r\sigma(b))) F_x'(\sigma(b)) [q\sigma'(b)] = 0. \quad (A.15)$$

As we have assumed that x is uniformly distributed, we have $F(x) = x$, and therefore

$$\sigma'(b) = \frac{1 - q\sigma(b)}{q[b - (1 + r\sigma(b))]}. \quad (A.16)$$

With the "separation of the variables" method this differential equation can be solved with the initial condition $\sigma(X) = 1$, i.e. imposing that the bidders with the highest refinancing costs post the highest bid, and assuming that $\sigma'(b) > 0$ we obtain for the inverse bidding function:

$$\sigma(b) = \frac{2(b-1)q - \sqrt{[2(b-1)q]^2 + 4r[2(b-q) - qr + 2X(q-1)]}}{2qr}. \quad (A.17)$$

The distribution function of the bids can be calculated as

$$\begin{aligned} F_j(b) &= F\left(b_j^{-1}(b)\right) = b_j^{-1}(b) \quad (A.18) \\ &= \sigma(b) \text{ for } b \in \left[q + X - qX + \frac{1}{2}qr, X\right]. \end{aligned}$$

With a Taylor expansion of this term it can be directly shown that for $r \to 0$, this converges pointwise on the definition range to the distribution function of the common value auction. We proceed indirectly by inverting, which gives us the bidding function:

$$b(x) = \frac{2(1-q)X - q(1-x)(2 + r + rx)}{2qx - 2}. \quad (A.19)$$

Now it is easy to see that for $r \to 0$, the bidding functions converge to

$$b(x) = \frac{(1-q)X - q(1-x)}{qx - 1}. \quad (A.20)$$

The inverse of $b(x)$ is

$$b_j^{-1}(b) = \frac{1}{q}\left(1 - \frac{(1-q)(X-1)}{b-1}\right). \tag{A.21}$$

This is exactly the distribution function for the mixed equilibrium of the pure common value auction. ■

Appendix B

Proofs for Chapter 3

Proof of Proposition 11 and 14

Without restriction of generality, we assume $\underline{c} = \underline{p} = 0$. We therefore have $\overline{c} = \Delta c$ and $\overline{p} = \Delta p$. The integrations will be immediately carried out for the general case with n outside bidders. The probability that the private values of all outside bidders are smaller than x is $\left(\frac{x}{\overline{p}}\right)^n$. Therefore the density function of the outside bidder with the highest private value is $g_n(x) = n\frac{x}{\overline{p}^n}^{n-1}$. Due to the different strategies for the outside bidders we have to distinguish the two main cases $\overline{c} > \overline{p}$ and $\overline{c} \leq \overline{p}$.

Case $\overline{c} > \overline{p}$

We have to further distinguish three cases:

1.If $p_{ins} + \hat{c} < \overline{p}$, then only the first of the formulas for the outsiders bidding function applies. If p_{out} is the private value of the outsider with the highest evaluation, the insider wins the auction iff $\pi_{out}(p_{out}) < \pi_{ins}(p_{ins}, \hat{c})$, i.e. iff $p_{out} \leq \frac{1}{2}(p_{ins} + \hat{c})$. In this case the dropout price of the outsider is $\pi_{out}(p_{out}) = 2p_{out}$. So the profit as a function of p_{out}, p_{ins} and $\hat{c}$ can be written as

$$\Pi_{ins}(\hat{c}, p_{ins}, p_{out}) = \begin{cases} \hat{c} + p_{ins} - 2p_{out}, & \text{if } p_{out} \leq \frac{1}{2}(p_{ins} + \hat{c}), \\ 0, & \text{if } p_{out} > \frac{1}{2}(p_{ins} + \hat{c}). \end{cases} \quad \text{(B.1)}$$

Taking the expectation with respect to p_{out} yields:

$$\Pi_{ins}(\hat{c}, p_{ins}) = \frac{1}{\overline{p}} \int_0^{\frac{1}{2}(p_{ins}+\hat{c})} (\hat{c} + p_{ins} - 2p_{out})\, g_n(p_{out})\, dp_{out} = \quad \text{(B.2)}$$

$$= \frac{(\hat{c} + p_{ins})^{1+n}}{2^n \overline{p}(1+n)}.$$

2.If $\overline{c} > p_{ins}+\hat{c} > \overline{p}$, then the insider wins the auction if $p_{out} \leq \frac{1}{2}\overline{p}$. The dropout price of the outsider can be determined with the same formula as before.

$$\Pi_{ins}\left(\hat{c}, p_{ins}, p_{out}\right) = \begin{cases} \hat{c} + p_{ins} - 2p_{out}, & \text{if } p_{out} \leq \frac{1}{2}\overline{p}, \\ 0, & \text{if } p_{out} > \frac{1}{2}\overline{p}. \end{cases} \tag{B.3}$$

The expected profit is then

$$\begin{aligned} \Pi_{ins}\left(\hat{c}, p_{ins}\right) &= \frac{1}{\overline{p}} \int_0^{\frac{1}{2}\overline{p}} \left(\hat{c} + p_{ins} - 2p_{out}\right) g_n\left(p_{out}\right) dp_{out} = \\ &= \frac{(1+n)\left(\hat{c} + p_{ins}\right) - n\overline{p}}{2^n\left(1+n\right)}. \end{aligned} \tag{B.4}$$

3. If $\overline{c} < p_{ins} + c$, then the insider wins if $p_{out} < \frac{1}{2}\left(\overline{p} + p_{ins} + \hat{c} - \overline{c}\right)$, i.e. the outsider drops out at $2p_{out} - \overline{p} + \overline{c}$, if $\frac{1}{2}\overline{p} \leq p_{out}$ and at the same price as before if $\frac{1}{2}\overline{p} > p_{out}$.

$$\Pi_{ins}\left(\hat{c}, p_{ins}, p_{out}\right) = \begin{cases} 0, & \text{if } p_{out} > \frac{1}{2}\left(\overline{p} + p_{ins} + \hat{c} - \overline{c}\right) \\ \hat{c} + p_{ins} - 2p_{out} + \overline{p} - \overline{c}, & \text{if } \frac{1}{2}\overline{p} < p_{out} \leq \frac{1}{2}\left(\overline{p} + p_{ins} + \hat{c} - \overline{c}\right) \\ \hat{c} + p_{ins} - 2p_{out} & \text{if } p_{out} \leq \frac{1}{2}\overline{p} \end{cases} \tag{B.5}$$

Integrating yields

$$\begin{aligned} \Pi_{ins}\left(\hat{c}, p_{ins}\right) &= \frac{1}{\overline{p}} \int_0^{\frac{1}{2}\overline{p}} \left(\hat{c} + p_{ins} - 2p_{out}\right) g_n\left(p_{out}\right) dp_{out} \\ &\quad + \frac{1}{\overline{p}} \int_{\frac{1}{2}\overline{p}}^{\frac{1}{2}\left(\overline{p}+p_{ins}+\hat{c}-\overline{c}\right)} \left(\hat{c} + p_{ins} - 2p_{out} + \overline{p} - \overline{c}\right) g_n\left(p_{out}\right) dp_{out} \\ &= \frac{\overline{c} - \overline{p}}{2^n} + \frac{\left(\hat{c} + p_{ins} - \overline{c} + \overline{p}\right)^{1+n}}{2^n\overline{p}^n\left(1+n\right)} \end{aligned} \tag{B.6}$$

The overall expected profit can now be calculated by integrating over the different domains of $\hat{c}$ and p_{ins} as:

$$\begin{aligned} \Pi_{ins}\left(\hat{c}, p\right) &= \frac{1}{\overline{pc}} \int_{\underline{p}}^{\overline{p}} \int_{\underline{c}}^{\overline{c}} \Pi_{ins}\left(\hat{c}, p_{ins}\right) d\hat{c}dp_{ins} \\ &= \frac{1}{\overline{pc}} \int_{\underline{p}}^{\overline{p}} \int_{\overline{c}+\underline{p}-p_{ins}} \frac{\left(\hat{c} + p_{ins}\right)^{1+n}}{2^n\overline{p}\left(1+n\right)} d\hat{c}dp_{ins} \\ &\quad + \frac{1}{\overline{pc}} \int_{\underline{p}}^{\overline{p}} \int_{\underline{c}+\overline{p}-p_{ins}}^{\overline{c}+\underline{p}-p_{ins}} \frac{(1+n)\left(\hat{c} + p_{ins}\right) - n\overline{p}}{2^n\left(1+n\right)} d\hat{c}dp_{ins} \\ &\quad + \frac{1}{\overline{pc}} \int_{\underline{p}}^{\overline{p}} \int_{\underline{c}}^{\underline{c}+\overline{p}-p_{ins}} \left[\frac{\overline{c} - \overline{p}}{2^n} + \frac{\left(\hat{c} + p_{ins} - \overline{c} + \overline{p}\right)^{1+n}}{2^n\overline{p}^n\left(1+n\right)}\right] d\hat{c}dp_{ins} \\ &= \frac{\overline{p}^2}{\overline{c}} \frac{8 - 2^{-n}\left(8 + n\left(5+n\right)\right)}{(n+1)\left(n+2\right)\left(n+3\right)} + \frac{\overline{c}}{2^{n+1}} - \frac{\overline{p}}{2^{n+1}} \frac{n-1}{n+1}. \end{aligned} \tag{B.7}$$

Substituting $\Delta c := \overline{c}$, $\Delta p := \overline{p}$ and setting $n = 1$ gives:

$$\Pi_{ins}(c,p) = \frac{\Delta p^2 + 6\Delta c^2}{24\Delta c}. \tag{B.8}$$

Case $\overline{c} < \overline{p}$

1.If $p_{ins} + \hat{c} < \overline{c}$, then we have the same situation as in the first case of the previous section:

$$\Pi_{ins}(\hat{c}, p_{ins}) = \frac{(\hat{c} + p_{ins})^{1+n}}{2^n \overline{p}(1+n)}. \tag{B.9}$$

2.If$\overline{c} < p_{ins} + \hat{c} < \overline{p}$, we have to apply the intermediary part of the outsider's bidding strategy. This means the insider wins if $p_{out} \leq \hat{c} + p_{ins} - \frac{1}{2}\overline{c}$.

$$\Pi_{ins}(\hat{c}, p_{ins}, p_{out}) = \begin{cases} 0, & \text{if } p_{out} > \hat{c} + p_{ins} - \frac{1}{2}(\underline{c} + \overline{c}), \\ \hat{c} + p_{ins} - p_{out} - \frac{1}{2}\overline{c} & \text{if } \frac{1}{2}\overline{c} \leq p_{out} \leq \hat{c} + p_{ins} - \frac{1}{2}\overline{c}, \\ \hat{c} + p_{ins} - 2p_{out} & \text{if } p_{out} \leq \frac{1}{2}\overline{p}. \end{cases} \tag{B.10}$$

The expected profit is then

$$\begin{aligned} \Pi_{ins}(\hat{c}, p_{ins}) &= \frac{1}{\overline{p}} \int_0^{\frac{1}{2}\overline{c}} (\hat{c} + p_{ins} - 2p_{out})\, g_n(p_{out})\, dp_{out} + \\ & \frac{1}{\overline{p}} \int_{\frac{1}{2}\overline{c}}^{\hat{c}+p_{ins}-\frac{1}{2}\overline{c}} \left(\hat{c} + p_{ins} - p_{out} - \frac{1}{2}\overline{c}\right) g_n(p_{out})\, dp_{out} \\ &= \frac{\overline{c}^{1+n} + (2\hat{c} + 2p_{ins} - \overline{c})^{n+1}}{2^{n+1}\overline{p}^n (1+n)} \end{aligned} \tag{B.11}$$

3.If $\overline{p}+ \ \underline{c} < p_{ins} + \hat{c}$, then the last part of the outsider's bidding function is relevant for deciding who wins the auction, but all three parts contribute to the determination of output prices:

$$\Pi_{ins}(\hat{c}, p_{ins}, p_{out}) = \begin{cases} 0, & \text{if } p_{out} > \frac{1}{2}(\overline{p} + p_{ins} + \hat{c} - \overline{c}), \\ \hat{c} + p_{ins} - 2p_{out} + \overline{p} - \overline{c}, & \text{if } \hat{c} + p_{ins} - \frac{1}{2}\overline{c} \leq p_{out} \\ & \leq \frac{1}{2}(\overline{p} + p_{ins} + \hat{c} - \overline{c}), \\ \hat{c} + p_{ins} - p_{out} - \frac{1}{2}\overline{c} & \text{if } \frac{1}{2}\overline{c} \leq p_{out} \leq \hat{c} + p_{ins} - \frac{1}{2}\overline{c}, \\ \hat{c} + p_{ins} - 2p_{out} & \text{if } p_{out} \leq \frac{1}{2}\overline{c}. \end{cases} \tag{B.12}$$

$$\begin{aligned}\Pi_{ins}(\hat{c},p_{ins}) &= \frac{1}{\overline{p}}\int_0^{\frac{1}{2}\overline{c}}\left(\hat{c}+p_{ins}-2p_{out}\right)g_n\left(p_{out}\right)dp_{out} \\ &\quad+\frac{1}{\overline{p}}\int_{\frac{1}{2}\overline{c}}^{\hat{c}+p_{ins}-\frac{1}{2}\overline{c}}\left(\hat{c}+p_{ins}-p_{out}-\frac{1}{2}\overline{c}\right)g_n\left(p_{out}\right)dp_{out}+ \\ &\quad\frac{1}{\overline{p}}\int_{\hat{c}+p_{ins}-\frac{1}{2}\overline{c}}^{\frac{1}{2}(\overline{p}+p_{ins}+\hat{c}-\overline{c})}\left(\hat{c}+p_{ins}-2p_{out}+\overline{p}-\overline{c}\right)g_n\left(p_{out}\right)dp_{out} \\ &= \frac{\overline{c}^{1+n}+2\left(\hat{c}+p_{ins}-\overline{c}+\overline{p}\right)^{n+1}-\left(2\overline{p}-\overline{c}\right)^{n+1}}{2^{n+1}\overline{p}^n\left(1+n\right)}.\end{aligned} \tag{B.13}$$

The overall expected profit can be calculated by integrating over the different domains of $\hat{c}$ and p_{ins} as

$$\begin{aligned}\Pi_{ins}(c,p) &= \frac{1}{\overline{pc}}\int_0^{\overline{c}}\int_0^{\overline{p}}\Pi_{ins}\left(\hat{c},p_{ins}\right)dp_{ins}d\hat{c}= \\ &= \frac{1}{\overline{pc}}\int_0^{\overline{c}}\int_0^{\overline{c}-\hat{c}}\frac{\left(\hat{c}+p_{ins}\right)^{1+n}}{2^n\overline{p}\left(1+n\right)}dp_{ins}d\hat{c} \\ &\quad+\frac{1}{\overline{pc}}\int_0^{\overline{c}}\int_{\overline{c}-\hat{c}}^{\overline{p}-\hat{c}}\frac{\overline{c}^{1+n}+\left(2\hat{c}+2p_{ins}-\overline{c}\right)^{n+1}}{2^{n+1}\overline{p}^n\left(1+n\right)}dp_{ins}d\hat{c} \\ &\quad+\frac{1}{\overline{pc}}\int_0^{\overline{c}}\int_{\overline{p}-\hat{c}}^{\overline{p}}\frac{\overline{c}^{1+n}+2\left(\hat{c}+p_{ins}-\overline{c}+\overline{p}\right)^{n+1}-\left(2\overline{p}-\overline{c}\right)^{n+1}}{2^{n+1}\overline{p}^n\left(1+n\right)}dp_{ins}d\hat{c} \\ &= \frac{1}{2^{n+2}\overline{p}^{n+1}\overline{c}\left(n+1\right)\left(n+2\right)\left(n+3\right)}\left[2^{n+5}\overline{p}^{n+3}\right. \\ &\quad-\overline{c}^{n+2}\left(\overline{c}\left(1+n\right)^2-2\left(2+n\right)\left(3+n\right)\overline{p}\right) \\ &\quad\left.-\left(2\overline{p}-\overline{c}\right)^{1+n}\left(\overline{c}^2\left(1+n\right)^2+16\overline{p}^2+2\overline{cp}\left(1+3n\right)\right)\right].\end{aligned} \tag{B.14}$$

For $n=1$ this equals:

$$\Pi_{ins}(c,p)=\frac{4c^2p+4p^3-c^3}{24p^2} \tag{B.15}$$

■

Proof of Proposition 12

We only consider the case of one outside bank.

Case $\overline{c}>\overline{p}$ If the inside bank drops out the price π always equals the banks valuation of the credit, i.e. $\pi=p_{ins}+\hat{c}$. Thus if an outsiders with private value p_{out} wins the auction his profit is $p_{out}+\hat{c}-\pi=p_{out}-p_{ins}$. In order to obtain the outsiders profit as a function of its private value, we first integrate over p_{ins} and

1. If $p_{out} < \frac{1}{2}\overline{p}$, then the outsider wins whenever $2p_{out} > p_{ins} + \hat{c}$. Thus in this case his expected profit is

$$\Pi_{out}(p_{out}) = \frac{1}{\overline{pc}} \int_0^{2p_{out}} \int_0^{2p_{out}-\hat{c}} (p_{out} - p_{ins})\, dp_{ins} d\hat{c} = \frac{2}{3\overline{pc}} p_{out}^3. \tag{B.16}$$

2. If $p_{out} > \frac{1}{2}\overline{p}$, then the outsider wins if $2p_{out} - \overline{p} + \overline{c} > p_{ins} + \hat{c}$. Integrating this profit over the relevant domains of p_{ins} and $\hat{c}$ we find:

$$\begin{aligned} \Pi_{out}(p_{out}) &= \frac{1}{\overline{pc}} \int_{2p_{out}-\overline{p}}^{\overline{p}} \int_0^{2p_{out}-\overline{p}+\overline{c}-p} (p_{out} - p_{ins})\, d\hat{c} dp_{ins} \qquad \text{(B.17)} \\ &+ \frac{1}{\overline{pc}} \int_0^{2p_{out}-\overline{p}} \int_0^{\overline{c}} (p_{out} - p_{ins})\, d\hat{c} dp_{ins} \\ &= \frac{4(\overline{p} - p)^3 - 3\overline{pc}(\overline{p} - 2p_{out})}{6\overline{pc}}. \end{aligned}$$

Taking the expectation with respect to p_{out} we obtain:

$$\begin{aligned} \Pi_{out} &= \frac{1}{\overline{p}} \int_0^{\frac{1}{2}\overline{p}} \frac{2}{3\overline{pc}} p_{out}^3 dp_{out} + \qquad \text{(B.18)} \\ &+ \frac{1}{\overline{p}} \int_{\frac{1}{2}\overline{p}0}^{\overline{p}} \frac{4(\overline{p} - p)^3 - 3\overline{pc}(\overline{p} - 2p_{out})}{6\overline{pc}} dp_{out} \\ &= \frac{p(p + 6c)}{48c}. \end{aligned}$$

Case $\overline{c} < \overline{p}$

1.If $p_{out} < \frac{1}{2}\overline{c}$, then we are in the same situation as in the first subcase of the case $\overline{c} > \overline{p}$ above.

$$\Pi_{out}(p_{out}) = \frac{2}{3\overline{pc}} p_{out}^3$$

2.If $\frac{1}{2}\overline{c} < p_{out} < \overline{p} - \frac{1}{2}\overline{c}$, then the outsider wins if $p_{out} + \frac{1}{2}\overline{c} > p_{ins} + \hat{c}$. Integrating yields:

$$\begin{aligned} \Pi_{out}(p_{out}) &= \frac{1}{\overline{pc}} \int_0^{\overline{c}} \int_0^{p_{out}+\frac{1}{2}\overline{c}-\hat{c}} (p_{out} - p_{ins})\, dp_{ins} d\hat{c} \qquad \text{(B.19)} \\ &= \frac{12p_{out}^2 - \overline{c}^2}{24\overline{p}}. \end{aligned}$$

3.If $p_{out} > \overline{p} - \frac{1}{2}\overline{c}$, then the outsider wins if $2p_{out} - \overline{p} + \overline{c} > p_{ins} + \hat{c}$. We are in the same situation as in the second subcase of the case $\overline{c} > \overline{p}$.

$$\begin{aligned}\Pi_{out}(p_{out}) &= \frac{1}{\overline{pc}} \int_{2p_{out}-\overline{p}+\overline{c}}^{\overline{c}} \int_{0}^{2p_{out}-\overline{p}+\overline{c}-\hat{c}} (p_{out} - p_{ins})\, dp_{ins} d\hat{c} + \\ &\quad \frac{1}{\overline{pc}} \int_{0}^{2p_{out}-\overline{p}+\overline{c}} \int_{0}^{\overline{p}} (p_{out} - p_{ins})\, dp_{ins} d\hat{c} \\ &= \frac{4(\overline{p} - p_{out})^3 - 3\overline{pc}(\overline{p} - 2p_{out})}{6\overline{pc}}.\end{aligned} \tag{B.20}$$

Taking the expectation with respect to p_{out} we obtain:

$$\begin{aligned}\Pi_{out} &= \frac{1}{\overline{p}} \int_{0}^{\frac{1}{2}\overline{c}} \left(\frac{2}{3\overline{pc}} p_{out}^3\right) dp_{out} + \\ &\quad \frac{1}{\overline{p}} \int_{\frac{1}{2}\overline{c}}^{\overline{p}-\frac{1}{2}\overline{c}} \frac{12p_{out}^2 - \overline{c}^2}{24\overline{p}} dp_{out} + \\ &\quad \frac{1}{\overline{p}} \int_{\overline{p}-\frac{1}{2}\overline{c}}^{\overline{p}} \frac{4(\overline{p} - p_{out})^3 - 3\overline{pc}(\overline{p} - 2p_{out})}{6\overline{pc}} dp_{out}.\end{aligned} \tag{B.21}$$

$$\Rightarrow \Pi_{out}(p, c) = \frac{8p^3 + c^2(c - 2p)}{48p^2}. \tag{B.22}$$

■

Proof of Proposition 13 (Expected Price)

We know that when the outsider wins, the price equals the insider's valuation. In the other case the price is determined by the bidding strategy of the outsiders. We are integrating the expected price in the same way as the insider's profit.

Case $\overline{c} > \overline{p}$

1. If $p_{ins} + \hat{c} < \overline{c}$, then

$$\pi(\hat{c}, p_{ins}, p_{out}) = \begin{cases} p_{ins} + \hat{c}, & \text{if } p_{out} > \frac{1}{2}(p_{ins} + \hat{c}), \\ 2p_{out}, & \text{if } p_{out} \leq \frac{1}{2}(p_{ins} + \hat{c}). \end{cases} \tag{B.23}$$

We first integrate over p_{out} :

$$\begin{aligned}\pi(\hat{c}, p_{ins}) &= \frac{1}{\overline{p}} \left(\int_{0}^{\frac{1}{2}(p_{ins}+\hat{c})} 2p_{out} dp_{out} + \int_{\frac{1}{2}(p_{ins}+\hat{c})}^{\overline{p}} (p_{ins} + \hat{c})\, dp_{out} \right) \\ &= \frac{1}{4\overline{p}} (\hat{c} + p_{ins})(4\overline{p} - \hat{c} - p_{ins}).\end{aligned} \tag{B.24}$$

2.If $\overline{c} > p_{ins} + \hat{c} > \overline{p},$ then

$$\pi\left(\hat{c}, p_{ins}, p_{out}\right) = \begin{cases} p_{ins} + \hat{c}, & \text{if } p_{out} > \frac{1}{2}\overline{p}, \\ 2p_{out}, & p_{out} \leq \frac{1}{2}\overline{p}. \end{cases} \tag{B.25}$$

The expected price is then

$$\begin{aligned} \pi\left(\hat{c}, p_{ins}\right) &= \frac{1}{\overline{p}}\left(\int_0^{\frac{1}{2}\overline{p}} 2p_{out}dp_{out} + \int_{\frac{1}{2}\overline{p}}^{\overline{p}} \left(p_{ins} + \hat{c}\right) dp_{out}\right) = \\ &= \frac{1}{4}\left[2\left(\hat{c} + p_{ins}\right) + \overline{p}\right]. \end{aligned} \tag{B.26}$$

3.If $\overline{p} + \underline{c} < p_{ins} + \hat{c},$ then

$$\pi\left(\hat{c}, p_{ins}, p_{out}\right) = \begin{cases} p_{ins} + \hat{c}, & \text{if } p_{out} > \frac{1}{2}\left(\overline{p} + p_{ins} + \hat{c} - \overline{c}\right), \\ 2p_{out} - \overline{p} + \overline{c}, & \text{if } \frac{1}{2}\overline{p} \leq p_{out} \leq \frac{1}{2}\left(\overline{p} + p_{ins} + \hat{c} - \overline{c}\right), \\ 2p_{out}, & \text{if } p_{out} \leq \frac{1}{2}\overline{p}. \end{cases} \tag{B.27}$$

Taking the expectation with respect to p_{out} we obtain:

$$\begin{aligned} \pi\left(\hat{c}, p_{ins}\right) &= \frac{1}{\overline{p}}\left(\int_0^{\frac{1}{2}\overline{p}} 2p_{out}dp_{out} + \int_{\frac{1}{2}\overline{p}}^{\frac{1}{2}\left(\overline{p}+p_{ins}+\hat{c}-\overline{c}\right)} \left(2p_{out} - \overline{p} + \overline{c}\right) dp_{out}\right) \\ &\quad + \frac{1}{\overline{p}}\int_{\frac{1}{2}\left(\overline{p}+p_{ins}+\hat{c}-\overline{c}\right)}^{\overline{p}} \left(p_{ins} + \hat{c}\right) dp_{out} \\ &= \frac{1}{4\overline{p}}\left[2\left(p_{ins} + \hat{c}\right)\overline{p} - \left(p_{ins} + \hat{c} - \overline{c}\right)^2 + \overline{p}^2\right]. \end{aligned} \tag{B.28}$$

Now we can integrate over p_{ins} and $\hat{c}$:

$$\frac{1}{\overline{pc}}\int_0^{\overline{c}}\int_0^{\overline{p}} \pi\left(\hat{c}, p_{ins}\right) dp_{ins}d\hat{c} = \tag{B.29}$$

$$\begin{aligned} = \frac{1}{\overline{pc}}\int_0^{\overline{p}}\int_0^{\overline{p}-\hat{c}} \frac{1}{4\overline{p}}\left(\hat{c} + p_{ins}\right)\left(4\overline{p} - \hat{c} - p_{ins}\right) dp_{ins}d\hat{c} \\ + \frac{1}{\overline{pc}}\int_0^{\overline{p}}\int_{\overline{p}-p_{ins}}^{\overline{c}-p_{ins}} \frac{1}{4}\left[2\left(\hat{c} + p_{ins}\right) + \overline{p}\right] d\hat{c}dp_{ins} \\ + \frac{1}{\overline{pc}}\int_{\overline{c}-\overline{p}}^{\overline{c}}\int_{\overline{c}-\hat{c}}^{\overline{p}} \frac{1}{4\overline{p}}\left[2\left(p_{ins} + \hat{c}\right)\overline{p} - \left(p_{ins} + \hat{c} - \overline{c}\right)^2 + \overline{p}^2\right] dp_{ins}d\hat{c} \end{aligned} \tag{B.30}$$

$$= \frac{1}{24}\left(6\overline{c} + 12\overline{p} - \frac{\overline{p}^2}{\overline{c}}\right). \tag{B.31}$$

Case $\overline{c} < \overline{p}$

1.If $p_{ins} + \hat{c} < \overline{p}$, then we are in the same situation as before and

$$\pi(\hat{c}, p_{ins}) = \frac{1}{4\overline{p}}(\hat{c} + p_{ins})(4\overline{p} - \hat{c} - p_{ins}). \tag{B.32}$$

2.If $\overline{c} < p_{ins} + \hat{c} < \overline{p}$, then

$$\pi(\hat{c}, p_{ins}, p_{out}) = \begin{cases} p_{ins} + \hat{c}, & \text{if } p_{ins} + \hat{c} - \frac{1}{2}\overline{c} \leq p_{out}, \\ p_{out} + \frac{1}{2}\overline{c}, & \text{if } p_{ins} + \hat{c} - \frac{1}{2}\overline{c} > p_{out} > \frac{1}{2}\overline{c}, \\ 2p_{out}, & \text{if } p_{out} \leq \frac{1}{2}\overline{c}. \end{cases} \tag{B.33}$$

The expected price is then

$$\begin{aligned} \pi(\hat{c}, p_{ins}) &= \frac{1}{\overline{p}} \int_0^{\frac{1}{2}\overline{c}} 2p_{out}dp_{out} \qquad \text{(B.34)} \\ &+ \frac{1}{\overline{p}} \int_{\frac{1}{2}\overline{c}}^{p_{ins}+\hat{c}-\frac{1}{2}\overline{c}} \left(p_{out} + \frac{1}{2}\overline{c}\right) dp_{out} + \frac{1}{\overline{p}} \int_{p_{ins}+\hat{c}-\frac{1}{2}\overline{c}}^{\overline{p}} (p_{ins} + \hat{c})\, dp_{out} \\ &= (p_{ins} + \hat{c}) - \frac{1}{4\overline{p}}\left[2\hat{c}^2 - 2\hat{c}\overline{c} + \overline{c}^2 + 4\hat{c}p_{ins} - 2\overline{c}p_{ins} + 2p_{ins}^2\right]. \end{aligned}$$

3.If $\underline{p} + \overline{c} < p_{ins} + \hat{c}$, then

$$\pi(\hat{c}, p_{ins}, p_{out}) = \begin{cases} p_{ins} + \hat{c} & \text{if } p_{ins} + \hat{c} - \frac{1}{2}\overline{c} < p_{out}, \\ 2p_{out} - \overline{p} + \overline{c} & \text{if } p_{ins} + \hat{c} - \frac{1}{2}\overline{c} > p_{out} > \overline{p} - \frac{1}{2}\overline{c}, \\ p_{out} + \frac{1}{2}\overline{c} & \text{if } \overline{p} - \frac{1}{2}\overline{c} > p_{out} > \frac{1}{2}\overline{c}, \\ 2p_{out} & \text{if } p_{out} \leq \frac{1}{2}\overline{c}. \end{cases} \tag{B.35}$$

Taking the expectation with respect to p_{out} we obtain:

$$\pi(\hat{c}, p_{ins}) = \frac{1}{4\overline{p}}\left[2(p_{ins} + \hat{c})\overline{p} - (p_{ins} + \hat{c} - \overline{c})^2 + \overline{p}^2\right]. \tag{B.36}$$

If we assume that $\underline{c} + \overline{p} - \underline{p} > \overline{c} + \underline{p} - \overline{p}$ the overall expected price can be calculated by integrating over the different domains of $\hat{c}$ and p_{ins} as

$$\frac{1}{\overline{pc}} \int_0^{\overline{c}} \int_0^{\overline{p}} \pi(\hat{c}, p_{ins})\, dp_{ins}d\hat{c} = \tag{B.37}$$

$$\begin{aligned} &= \frac{1}{\overline{p}^2\overline{c}} \int_0^{\overline{c}} \int_0^{\overline{c}+\underline{p}-\hat{c}} \frac{1}{4\overline{p}}(\hat{c} + p_{ins})(4\overline{p} - \hat{c} - p_{ins})\, dp_{ins}d\hat{c} \qquad \text{(B.38)} \\ &+ \frac{1}{\overline{p}^2\overline{c}} \int_0^{\overline{c}} \int_{\overline{c}-\hat{c}}^{\overline{p}-c} (p_{ins} + \hat{c}) - \\ &\qquad \frac{1}{4\overline{p}}\left[2\hat{c}^2 - 2\hat{c}\overline{c} + \overline{c}^2 + 4\hat{c}p_{ins} - 2\overline{c}p_{ins} + 2p_{ins}^2\right] dp_{ins}d\hat{c} \\ &+ \frac{1}{\overline{p}^2\overline{c}} \int_0^{\overline{c}} \int_{\overline{p}-\hat{c}}^{\overline{p}} \frac{1}{4\overline{p}}\left[2(p_{ins} + \hat{c})\overline{p} - (p_{ins} + \hat{c} - \overline{c})^2 + \overline{p}^2\right] dp_{ins}d\hat{c} \end{aligned}$$

$$= \frac{\bar{c}^3 - 4\bar{c}^2\bar{p} + 12\bar{c}\bar{p}^2 + 8\bar{p}}{24\bar{p}^2}. \tag{B.39}$$

■

Proof of Corollary 5

The profit in an auction with $n+1$ identically informed bidders whose private values are uniformly distributed on an interval with length Δp is $\frac{\Delta p}{(n+1)(n+2)}$. It has to be shown that for sufficiently high n this is higher than the profit as an insider.

For $\Delta c \leq \Delta p$, it has to be shown that

$$\begin{aligned} &\frac{1}{2^{n+2}\Delta p^{n+1}\Delta c\,(n+1)\,(n+2)\,(n+3)}\Big[2^{n+5}\Delta p^{n+3} \\ &-\Delta c^{n+2}(\Delta c\,(1+n)^2 - 2\,(2+n)\,(3+n)\,\Delta p) - (2\Delta p - \Delta c)^{1+n} \\ &\left(\Delta c^2\,(1+n)^2 + 16\Delta p^2 + 2\Delta c\Delta p\,(1+3n)\right)\Big] - \frac{\Delta p}{(n+1)\,(n+2)} < 0\,. \end{aligned} \tag{B.40}$$

Multiplying with $4\Delta p\,(n+1)\,(n+2)$ yields

$$\Leftarrow \quad -4\Delta p^2 + \frac{1}{\Delta c\Delta p^n 2^n\,(3+n)}\left[2^{n+5}\Delta p^{3+n} - \Delta c^{2+n}\left(\Delta c\,(1+n)^2\right.\right. \tag{B.41}$$

$$\begin{aligned} &\left.-2\Delta p\,(2+n)\,(3+n)\right) - (2\Delta p - \Delta c)^{1+n} \\ &\left.\left(\Delta c\,(1+n)^2 + 16\Delta p^2 + 2\Delta c\Delta p\,(1+3n)\right)\right] < 0. \end{aligned} \tag{B.42}$$

Since $\Delta p > \Delta c$, the second term converges to zero for $n \to \infty$,. hence the whole term converges against $-4\Delta p^2 < 0$.

For $\Delta c > \Delta p$:

$$\frac{\Delta p^2}{\Delta c}\frac{8 - 2^{-n}\,(8 + n\,(5+n))}{(n+1)\,(n+2)\,(n+3)} + \frac{\Delta c}{2^{n+1}} - \frac{\Delta p}{2^{n+1}}\frac{n-1}{n+1} - \frac{\Delta p}{(n+1)\,(n+2)} < 0. \tag{B.43}$$

Multiplying with $\Delta c\,(n+1)\,(n+2)\,(n+3)\,2^{1+n}$ yields:

$$\begin{aligned} &\Delta c^2\,(n+1)\,(n+2)\,(n+3) - \\ &\Delta c\Delta p\,(3+n)\left(2^{1+n} - 2 + n + n^2\right) + \\ &\Delta p^2 2\,(8\,(2^n - 1) - n\,(5+n)) < 0. \end{aligned} \tag{B.44}$$

As exponential functions grow faster than any polynomial, this converges asymptotically to $-\Delta c\Delta p\,(3+n)\,2^{1+n} + \Delta p^2 2^{n+5} < 0$, which can be simplified to $-\Delta c\,(3+n) + \Delta p 2^4 < 0$. his is true for big enough n ■

Appendix C

Proofs for Chapter 4

Existence and Uniqueness

Three preliminary properties of any equilibrium will be derived before existence and uniqueness can be proved:

Lemma 1 *The bidding supports of both bidders have to be identical*

Proof. Suppose one of the bidders is bidding in a open region where the competitor does not place bids. Then the bidder cannot be indifferent between these bids. Bidding higher would increase the profit in case he wins and not decrease his probability of winning. ■

Lemma 2 *The bidding supports have to be single intervals including the highest possible bid.*

Proof. Suppose there is a open region U in [1,X] in which the bidders place no bids, but below which the bidders place bids. Define $x := \sup\{z|z < U \wedge z \in Support\}$. Then the profit for bidding in U must be bigger than bidding x : The interest rate rises, but the winning probability does not change. Hence the support has to be a single interval including the highest possible bid. ■

Lemma 3 *One of the distribution functions may be discontinuous at the highest possible bid. Elsewhere the distribution functions are continuous. On the open support they are differentiable.*

Proof. Integrating the equilibrium condition 4.1 by parts and rearranging the terms yields:

$$\begin{aligned} K &= (b-1)\,q\,(1-F_i(b)) + (b-1)\,(1-q) \\ &\quad + \alpha_j q F_i(b)\,(X-b) - \alpha_j q F_i(1)\,(X-1) \\ &\quad + \alpha_j \int_1^b qF_i(x)\,dx + \alpha_j q\,(1-F_i(b))\,(X-b) + \alpha_j\,(1-q)\,(X-b)\,, \end{aligned} \tag{C.1}$$

$\Rightarrow$

$$F_i\left(b\right) = \frac{\alpha_j\left[q\left(F_i(1)(X-1)-(X-b)-\int_1^b F_i(x)dx\right)-(1-q)(X-b)\right]-(b-1)+K}{\left[\alpha_j q(X-b)-(b-1)q-\alpha_j q(X-b)\right]}. \tag{C.2}$$

The right hand side of this equation is differentiable, hence $F_i\left(b\right)$ is differentiable, too.

Proof of proposition 15

Let $[l, X]$ be the support of both bidder's distribution functions. In a mixed strategy equilibrium bank j must be indifferent between any bids on the open kernel of the support i.e. $\frac{\partial}{\partial b}\pi_j\left(b\right) = 0 \ \forall b \in \left(l, X\right), j = 1, 2.$

This condition leads to the following differential equations:

$$\left(\alpha_j - 1\right)\left[qF_i\left(b\right) - 1\right] - \left(b-1\right)qF_i'\left(b\right) = 0, \ i = 1, 2. \tag{C.3}$$

$$\Rightarrow F_i'\left(b\right) = \frac{\left(1-\alpha_j\right)}{\left(b-1\right)}\left(\frac{1}{q} - F_i\left(b\right)\right), \ i = 1, 2. \tag{C.4}$$

Since on the open support, the distribution functions are differentiable they have to be there a solution of this differential equation. One of the distribution functions has also to be continuous at $b = X$, Hence it can be obtained as the unique solution of C.4 for the boundary value $F\left(X\right) = 1$:

$$F_i\left(b\right) = \frac{1}{q} - \frac{1-q}{q}\left(\frac{X-1}{b-1}\right)^{1-\alpha_j}. \tag{C.5}$$

F_i is nonzero for

$$b \in \left[1 + \left(X-1\right)\left(1-q\right)^{\frac{1}{1-\alpha_j}}, X\right]. \tag{C.6}$$

From C.4 it can be seen that for higher α_j the distribution function of bidder i becomes flatter. Both bidders have to randomize on the same support and only one distribution function may be discontinuous at the highest bid. As $\alpha_2 > \alpha_1$ this must be F_1. Sine his bidding function has to be zero at the lower bound of the support this can be calculated by solving equation C.4 with α_2 and the initial condition $F_1\left(1 + \left(X-1\right)\left(1-q\right)^{\frac{1}{1-\alpha_2}}\right) = 0$.

The solution of this differential equation is:

$$\begin{aligned} F_1\left(b\right) &= \frac{1}{q} - \frac{1}{q}\left(\frac{\left(1-q\right)^{\frac{1}{1-\alpha_1}}\left(X-1\right)}{b-1}\right)^{1-\alpha_2} \quad \text{for } b \in \left[1 + \left(X-1\right)\left(1-q\right)^{\frac{1}{1-\alpha_1}}, X\right), \\ &= 1 \text{ for } b \geqq X, \\ &= 0 \text{ elswhere.} \end{aligned} \tag{C.7}$$

The bidding function of the second bank can be obtained with $i = 2$, $j = 1$ form C.5

$$\begin{aligned} F_2(b) &= \frac{1}{q} - \frac{1}{q}\frac{(1-q)(X-1)^{1-\alpha_1}}{(b-1)^{1-\alpha_1}} \text{for } b \in \left[1+(X-1)(1-q)^{\frac{1}{1-\alpha_1}}, X\right], \\ &= 1 \text{ for } b \geqq X, \qquad \text{(C.8)} \\ &= 0 \text{ elswhere.} \end{aligned}$$

■

Comparative statics

The profit of bank i on the credit business can be calculated as

$$\pi_i^C(b) = q\lambda(b-1)[q(1-F_j(b)) + 1 - q] = q\lambda(b-1)[1 - qF_j(b)]. \qquad \text{(C.9)}$$

The overall profit of a bank is independent of b, but not the profit on the credit business. Hence this has to be integrated over the equilibrium distribution of bank $i's$ bids in order to obtain the expected profit.

For bank 1 the result is:

$$\begin{aligned} \Pi_1^C &= \int_{1+(X-1)(1-q)^{\frac{1}{1-\alpha_1}}}^{X} \pi_1^C(b)\, dF_1(b) \qquad \text{(C.10)} \\ &= \frac{\lambda(X-1)(1-q)^2}{1-\alpha_1-\alpha_2}\left[\alpha_1\left(1-(1-q)^{\frac{\alpha_2-\alpha_1}{1-\alpha_1}}\right) - (1-\alpha_2)\left(1-(1-q)^{\frac{2\alpha_1-1}{1-\alpha_1}}\right)\right]. \end{aligned}$$

The expected profit of bank 2's credit department is

$$\begin{aligned} \Pi_2^C &= \int_{1+(X-1)(1-q)^{\frac{1}{1-\alpha_1}}}^{X} \pi_2^C(b)\, dF_2(b) \qquad \text{(C.11)} \\ &= \frac{\lambda(X-1)(1-\alpha_1)}{1-\alpha_1-\alpha_2}\left[(1-q)^{\frac{1}{1-\alpha_1}}\left(1-(1-q)^{\frac{1-\alpha 1-\alpha 2}{1-\alpha_1}}\right)\right]. \end{aligned}$$

Proof of proposition 16

The following proofs are always based on the inequality $\ln(x) < x - 1 \Leftrightarrow \ln(z) > 1 - 1/z$ for $x, z > 0$.

The derivative of C.10 with respect to α_2 is:

$$\begin{aligned} \frac{\partial}{\partial \alpha_2}\Pi_1^C &= \alpha_1 \frac{\lambda(X-1)(1-q)^{\frac{2-\alpha 1-\alpha 2}{1-\alpha_1}}}{(1-\alpha_1-\alpha_2)^2}\left[(1-q)^{-\frac{1-\alpha 1-\alpha 2}{1-\alpha_1}} - 1\right. \\ &\quad \left. + \ln\left((1-q)^{\frac{1-\alpha 1-\alpha 2}{1-\alpha_1}}\right)\right] > 0. \qquad \text{(C.12)} \end{aligned}$$

All the factors are positive, if $\alpha_1 > 0$ they are strictly positive.

The derivative of C.11 with respect to α_2 is:

$$\frac{\partial}{\partial \alpha_2}\Pi_2^C = (1-\alpha_1)\frac{\lambda(X-1)(1-q)^{\frac{2-\alpha 1-\alpha 2}{1-\alpha_1}}}{(1-\alpha_1-\alpha_2)^2}\left[(1-q)^{-\frac{1-\alpha 1-\alpha 2}{1-\alpha_1}} - 1 + \ln\left((1-q)^{\frac{1-\alpha 1-\alpha 2}{1-\alpha_1}}\right)\right] > 0. \tag{C.13}$$

■

Proof of proposition 17

The derivative of C.10 with respect to α_1 is:

$$\begin{aligned}
\frac{\partial}{\partial \alpha_1}\Pi_1^C &= \frac{(1-\alpha_2)}{(1-\alpha_1)}\frac{\lambda(X-1)(1-q)^{\frac{1}{1-\alpha_1}}}{(1-\alpha_1-\alpha_2)^2}\left[(1-\alpha_1)\left[1-(1-q)^{\frac{1-\alpha 1-\alpha 2}{1-\alpha_1}}\right]+\right.\\
&\quad \left.+\left(1-\alpha_1(1-q)^{\frac{1-\alpha 1-\alpha 2}{1-\alpha_1}}\right)\frac{(1-\alpha_1-\alpha_2)}{(1-\alpha 1)}\ln(1-q)\right] && \text{(C.14)}\\
&< K\left[(1-\alpha_1)\left[1-(1-q)^{\frac{1-\alpha 1-\alpha 2}{1-\alpha_1}}\right]-\right.\\
&\quad \left.\left[1-\alpha_1(1-q)^{\frac{1-\alpha 1-\alpha 2}{1-\alpha_1}}\right]\left(1-(1-q)^{\frac{1-\alpha 1-\alpha 2}{1-\alpha_1}}\right)\right] && \text{(C.15)}\\
&= K(-\alpha_1)\left(1-(1-q)^{\frac{1-\alpha 1-\alpha 2}{1-\alpha_1}}\right)^2 < 0.
\end{aligned}$$

we have used that $\alpha 1 + \alpha 2 < 1$.

The derivative of C.11 with respect to α_1 is:

$$\begin{aligned}
\frac{\partial}{\partial \alpha_1}\Pi_2^C &= \frac{\lambda(X-1)(1-q)^{\frac{1}{1-\alpha_1}}}{(1-\alpha_1-\alpha_2)^2}\left[\alpha_2\left(1-(1-q)^{\frac{1-\alpha 1-\alpha 2}{1-\alpha_1}}\right)+\right.\\
&\quad \left.+\left[1-(1-q)^{\frac{1-\alpha 1-\alpha 2}{1-\alpha_1}}+\alpha_2(1-q)^{\frac{1-\alpha 1-\alpha 2}{1-\alpha_1}}\right]\ln\left((1-q)^{\frac{1-\alpha 1-\alpha 2}{1-\alpha_1}}\right)\right]\\
&< K'\alpha_2\left[\left(1-(1-q)^{\frac{1-\alpha 1-\alpha 2}{1-\alpha_1}}\right)\right. && \text{(C.16)}\\
&\quad \left.-\left[1-(1-q)^{\frac{1-\alpha 1-\alpha 2}{1-\alpha_1}}+\alpha_2(1-q)^{\frac{1-\alpha 1-\alpha 2}{1-\alpha_1}}\right]\left(1-(1-q)^{\frac{1-\alpha 1-\alpha 2}{1-\alpha_1}}\right)\right]\\
&= K'\left(1-(1-q)^{\frac{1-\alpha 1-\alpha 2}{1-\alpha_1}}\right)\left[\alpha_2-1+(1-q)^{\frac{1-\alpha 1-\alpha 2}{1-\alpha_1}}-\alpha_2(1-q)^{\frac{1-\alpha 1-\alpha 2}{1-\alpha_1}}\right]\\
&= K'\left(1-(1-q)^{\frac{1-\alpha 1-\alpha 2}{1-\alpha_1}}\right)^2(\alpha_2-1) < 0.
\end{aligned}$$

■

Proof of proposition 18

For $\alpha_1 = \alpha_2 = \alpha$, the profit of both banks are identical, we can use any of the formulas C.10 or C.11.

$$\begin{aligned} \frac{\partial}{\partial \alpha}\Pi^C &= \frac{\partial}{\partial \alpha}\frac{\lambda (X-1)(1-\alpha)}{1-2\alpha}\left[(1-q)^{\frac{1}{1-\alpha}}\left(1-(1-q)^{\frac{1-2\alpha}{1-\alpha}}\right)\right] \qquad \text{(C.17)} \\ &= \frac{\lambda (X-1)(1-q)^{\frac{1}{1-\alpha}}}{(1-2\alpha)^2}\left[1-(1-q)^{\frac{1-2\alpha}{1-\alpha}}+\frac{1-2\alpha}{1-\alpha}\ln(1-q)\right] \\ &< 0. \end{aligned}$$

■

Appendix D

Proofs for Chapter 5

Proof of proposition 19

Proof. Proposition 19 is proven by giving a generic example of an economy in which credit rationing occurs.

Let $\alpha = \beta = \gamma = \frac{1}{3}$, $\lambda_t = 0.2$, $\lambda_u = 0.7$, $\hat{\rho} = 0.6$, $\pi < \frac{\Pi}{2}$. Suppose that the risk profile of the ecopnomy is such that, by rationing credit and only financing a fraction $\hat{x}$, the average risk in the sample can be reduced to $\lambda_t(\hat{x}) = 0.1$. It will be shown that none of the firms' strategies leads to an equilibrium if all firms are financed. The only possibility for honest as well as opportunistic banks to build up a good reputation and finance firms is to reduce the amount of riskiness among their clients by reducing x, i.e. by rationing the more risky firms.

Solving the game by backward induction, the firms' strategies in the last period have to be considered first. Two possible strategies are feasible for the firms in the second repetition of the game:

a) The firms only go to a bank with clean track record

b) The firms also go to a bank if they observe not more than one renegotiation.

c) The firms go to a bank in any case.

Case a) What is the optimal amount of exploitation $y \in [0, 1]$ of the opportunistic player, given that firms do not go to a bank that they know to have renegotiated a credit? If an opportunistic bank chooses to exploit a fraction y of all firms, a fraction $\lambda_t + (1 - \lambda_t) y$ of the firms has its credit contract renegotiated. With a probability of $[1 - (\lambda_t + (1 - \lambda_t) y)]^2$ no renegotiation of interest rates is observed in a randomly drawn sample of two firms. As there are no incentives to maintain a reputation after the game is finished an opportunistic bank will always exploit in the last period. Hence the two-period profit of an opportunistic bank can be calculated as:

$$\Pi y + \pi (1 - y) + [1 - (\lambda_t + (1 - \lambda_t) y)]^2 \Pi.$$

This function is obviously convex. It can therefore have no interior maximum, the maxima will be on the borders $y = 0$ or $y = 1$. If the bank exploits all customers its profit is Π. If the bank does not exploit any customer i.e. for $y = 0$, the profit is $\pi + (1-\lambda_t)^2 \Pi$. As $\pi = \frac{\Pi}{2}$, this can only be bigger than Π if .

$$\frac{\Pi}{2} + (1-\lambda_t)^2 \Pi > \Pi$$

$$\Rightarrow \lambda_t < \hat{\lambda}_1 = 1 - \sqrt{\frac{1}{2}} < 0.29.$$

For small λ_t the opportunistic players have no incentives to exploit any firms in the first period, they will perfectly mimic the honest types. For higher λ_t the opportunistic banks will immediately exploit all firms.

Given this behavior, is it optimal for firms to go to banks with a clean track record, but not to banks for which they have observed at least one negotiation of the interest rate? We first assume that all firms in the sample are financed i.e. that $\lambda_t = 0.2$.

No rationing:

Observing two non-renegotiated credit contracts, they believe that they will be exploited with probability:

$$\begin{aligned} P(\text{ex}|\text{clean}) &= \frac{\beta(1-\lambda_t)^2 + \gamma\lambda_u(1-\lambda_u)^2}{(\alpha+\beta)(1-\lambda_t)^2 + \gamma(1-\lambda_u)^2} \\ &= 0.52 < \hat{\rho}. \end{aligned}$$

Since this is smaller than $\hat{\rho}$ firms will always go to a bank with a clean track record.

The probability of being exploited by a bank with mixed records is:

$$\begin{aligned} P(\text{ex}|\text{mixed}) &= \frac{\beta\lambda_t(1-\lambda_t) + \gamma\lambda_u^2(1-\lambda_u)}{(\alpha+\beta)\lambda_t(1-\lambda_t) + \gamma\lambda_u(1-\lambda_u)} \\ &= 0.58 < \hat{\rho}. \end{aligned}$$

and the probability of beeeing exploited by a bank with bad record

$$\begin{aligned} P(\text{ex}|\text{bad}) &= \frac{\beta\lambda_t^2 + \gamma\lambda_u^2}{(\alpha+\beta)\lambda_t^2 + \gamma\lambda_u^2} \\ &= 0.67 > \hat{\rho}. \end{aligned}$$

The probability of being exploited by a bank with clean as well as with mixed track record is smaller than $\hat{\rho}$, whereas the probability of beeeing exploited by a bank with bad recored is higher than $\hat{\rho}$. In this case the firms will go to banks with clean and mixed records, which is contraditory to assumption a).

Rationing:

Rationing reduces the risk in the sample and increases the risk of beeing exploited after having obsered a mixed track record. The critical risk level can be calculated as

$$\begin{aligned} P(\text{ex}|\text{mixed}) &= \frac{\beta\lambda_t(1-\lambda_t)+\gamma\lambda_u^2(1-\lambda_u)}{(\alpha+\beta)\lambda_t(1-\lambda_t)+\gamma\lambda_u(1-\lambda_u)} = 0.6 \\ &\Rightarrow \lambda_t = 0.12. \end{aligned}$$

If a trustful bank reduces the risk of its sample to $\lambda_t(\hat{x}) = 0.12$, opportunists and usurers will have to follow. For $\lambda_t(\hat{x}) = 0.12$, firms will have no incentives to go to banks which have renegotiated and opportunists will have no incentives to exploit in the first period. Therefore this is an equilibrium.

Case b) It is assumed that the firms go to a bank if they observe not more than one renegotiation.

If firms think that in equilibrium they are only exploited by usurers, the opportunist's profit function is:

$$\Pi y + \pi(1-y) + \left[1 - (\lambda_t + (1-\lambda_t)y)^2\right]\Pi. \tag{D.1}$$

The profit for full exploitation is the same as in the previous case. However now the profit for complete honesty is $\pi + (1-\lambda_t^2)\Pi$. The profit function is now concave and has a maximum at

$$y = \frac{\Pi - \pi}{2\Pi(1-\lambda_t)^2} - \frac{\lambda_t}{(1-\lambda_t)} > 0.14. \tag{D.2}$$

Hence the opportunistic players will exploit a nonzero fraction of the firms. This is not consistent with the firms' believes. Is an interior equilibrium possible? The opportunistic players are now renegotiationg a fraction $\lambda_{opp} = \lambda_t + (1-\lambda_t)y = 0.31$ of the firms. Hence the probability of beeing exploited by a bank with mixed track recoprd is:

$$\begin{aligned} P(\text{ex}|\text{mixed}) &= \frac{\beta\lambda_{opp}(1-\lambda_{opp})+\gamma\lambda_u^2(1-\lambda_u)}{\alpha\lambda_t(1-\lambda_t)+\beta\lambda_{opp}(1-\lambda_{opp})+\gamma\lambda_u(1-\lambda_u)} \\ &= 0.61 > \hat{\rho}. \end{aligned}$$

and no firm will go to such a bank.

Case c) The probability of beeing exploited by a bank with a track recored of two renegotiations is:

$$\begin{aligned} P(\text{ex}|\text{bad}) &= \frac{\beta\lambda_t^2+\gamma\lambda_u^3}{(\alpha+\beta)\lambda_t^2+\gamma\lambda_u^2} \\ &= 0.67 > \hat{\rho}. \end{aligned}$$

As this is higher than $\hat{\rho}$ no firm will go to such a bank. ■

Proof of proposition 20

Given the assumption the risk profile is:

$$\lambda(\hat{x}) = \frac{1}{\hat{x}} \int_0^{\hat{x}} p(e^*, x)\, dx = \begin{cases} 0 & \text{for } \hat{x} < \xi, \\ \frac{p_2^2}{\hat{x}}(\hat{x} - \xi) & \text{for } \hat{x} \geq \xi. \end{cases} \tag{D.3}$$

We have

$$\lambda'(\hat{x}) = \begin{cases} 0 & \text{for } \hat{x} < \xi, \\ \xi \frac{p_2}{\hat{x}^2} & \text{for } \hat{x} \geq \xi. \end{cases} \tag{D.4}$$

The firm's profit is then

$$\Pi_{1+2}(x) = \begin{cases} \pi(x+1) & \text{for } \hat{x} < \xi, \\ \pi\left[1 + \left(1 - \frac{p_2^2}{\hat{x}}(\hat{x} - \xi)\right)^2\right] & \text{for } \hat{x} \geq \xi. \end{cases} \tag{D.5}$$

Obviously the function is convex on $(\xi, 1)$.Therefore the maximum has to be either at $x = \xi$ or at $x = 1$.

The interior maximum at $x = \xi$ corresponds to the credit rationing equilibrium. It is the overall maximum if:

$$\pi(\xi + 1) > \pi\left[1 + \left(1 - p_2^2(1 - \xi)\right)^2\right], \tag{D.6}$$

$$\Leftarrow p_2^2 > \frac{1}{1 + \sqrt{\xi}}. \tag{D.7}$$

■

Beiträge zur betriebswirtschaftlichen Forschung

Schriftenreihe herausgegeben von:

Prof. Dr. Dr. h. c. mult. Horst Albach, Bonn
Prof. Dr. Sönke Albers, Kiel
Prof. Dr. Dr. h. c. Herbert Hax, Köln
Prof. Dr. Bernhard Pellens, Bochum

Auswahl der erschienenen Bände:

Band 7 H. Albach
Wirtschaftlichkeitsrechnung bei unsicheren Erwartungen

Band 10 H. Hax
Vertikale Preisbindung in der Markenartikelindustrie

Band 14 D. Schneider
Die wirtschaftliche Nutzungsdauer von Anlagegütern

Band 40 B. Rudolph
Die Kreditgewährungsentscheidung der Banken

Band 43 A. Luhmer
Maschinelle Produktionsprozesse

Band 46 H. Simon
Preisstrategien für neue Produkte

Band 59
T. Hartmann-Wendels
Dividendenpolitik bei asymmetrischer Informationsverteilung

Band 61 R. Ewert
Rechnungslegung, Gläubigerschutz und Agencyprobleme

Band 65 W. Neus
Ökonomische Agency-Theorie und Kapitalgleichgewicht

Band 66 U. Backes-Gellner
Ökonomie der Hochschulforschung

Band 71 M. Steven
Produktion und Umweltschutz

Band 72 E. Terberger
Neo-institutionalistische Ansätze

Beiträge zur betriebswirtschaftlichen Forschung

Schriftenreihe herausgegeben von:

Prof. Dr. Dr. h. c. mult. Horst Albach, Bonn
Prof. Dr. Sönke Albers, Kiel
Prof. Dr. Dr. h. c. Herbert Hax, Köln
Prof. Dr. Bernhard Pellens, Bochum

Zuletzt erschienen:

Band 83 M. Pfennig
Optimale Steuerung des Währungsrisikos mit derivativen Instrumenten

Band 84 S. P.-B. Schiemann
Markt- und Organisationsstrukturen im Differenzierungswettbewerb

Band 85 E. Theissen
Organisationsformen des Wertpapierhandels

Band 86 M. Gürtler
Lebesguesche Optionspreistheorie

Band 87 R. F. Göx
Strategische Transferpreispolitik im Dyopol

Band 88 S. Wielenberg
Investitionen in Outsourcing-Beziehungen

Band 89 D. Köster
Wettbewerb in Netzproduktmärkten

Band 90 B. Skiera
Mengenbezogene Preisdifferenzierung bei Dienstleistungen

Band 91 A. Kempf
Wertpapierliquidität und Wertpapierpreise

Band 92 M. Uhrig-Homburg
Fremdkapitalkosten, Bonitätsrisiken und optimale Kapitalstruktur

Band 93 J. Budde
Effizienz betrieblicher Informationssysteme

Band 94 M. Tröge
Competition in Credit Markets

Band 95 H. Ernst
Erfolgsfaktoren neuer Produkte

GPSR Compliance

The European Union's (EU) General Product Safety Regulation (GPSR) is a set of rules that requires consumer products to be safe and our obligations to ensure this.

If you have any concerns about our products, you can contact us on ProductSafety@springernature.com

In case Publisher is established outside the EU, the EU authorized representative is:

Springer Nature Customer Service Center GmbH
Europaplatz 3
69115 Heidelberg, Germany

Batch number: 10725993

Printed by Printforce, the Netherlands